薪一传

Keep
The Fire Going

薪传
Keep The Fire Going

中国烹饪名家
The Chinese Culinary Masters

李启贵
Li Qi Gui

京菜
Beijing Cuisine

北京市商务局 编

中国轻工业出版社

图书在版编目（CIP）数据

中国烹饪名家. 李启贵：京菜 / 北京市商务局编. -- 北京：中国轻工业出版社，2021.8
ISBN 978-7-5184-3594-4

Ⅰ.①中… Ⅱ.①北… Ⅲ.①中式菜肴—烹饪 Ⅳ.①TS972.117

中国版本图书馆CIP数据核字(2021)第144626号

责任编辑：张 弘　　责任终审：张乃东　　整体设计：莱恩品牌
责任校对：晋 洁　　责任监印：张京华

出版发行：中国轻工业出版社（北京东长安街6号，邮编：100740）
印　　刷：鸿博昊天科技有限公司
经　　销：各地新华书店
版　　次：2021年8月第1版第1次印刷
开　　本：889×1194　1/16　印张：15.5
字　　数：250千字
书　　号：ISBN 978-7-5184-3594-4　定价：280.00元
邮购电话：010-65241695
发行电话：010-85119835　传真：85113293
网　　址：http://www.chlip.com.cn
Email：club@chlip.com.cn
如发现图书残缺请与我社邮购联系调换
210696S1X101ZBW

编委会

编委会主任：闫立刚

编委会副主任：孙 尧　刘梅英

执 行 主 编：闫立刚　孙 尧　云 程

编　　　辑：云 程　于 文　胡 平　李士靖　王文桥　陈连生　赵 珩
　　　　　　刘一达　侯玉瑞　王美萍

特 约 编 辑：宋子刚

策　　　划：李 东

文 字 撰 稿：周家旺

菜 品 撰 稿：李启贵

翻　　　译：北京精细艾华国际翻译有限公司

特 别 鸣 谢：北京烹饪协会

支 持 单 位：金丰集团

Editorial Committee

Director of Editorial Committee: Yan Ligang

Deputy Director of Editorial Committee: Sun Yao　Liu Meiying

Executive Editors: Yan Ligang　Sun Yao　Yun Cheng

Editors: Yun Cheng　Yu Wen　Hu Ping　Li Shijing　Wang Wenqiao　Chen Liansheng　Zhao Heng
　　　　Liu Yida　Hou Yurui　Wang Meiping

Contributing Editor: Song Zigang

Planner: Li Dong

Text Writer: Zhou Jiawang

Dish Writer: Li Qigui

Translator: Beijing Refined Aihua International Translation Co., Ltd.

Acknowledgement: Beijing Cuisine Association

Supporter: Jinfeng Group

李启贵

Li Qigui

2002年，第4届中国烹饪世界大赛在马来西亚举行，世界烹饪联合会创始会长张世尧为李启贵大师颁奖。
In 2002, the 4th World Championship of Chinese Cuisine was held in Malaysia, and Zhang Shiyao, the founding president of the World Association of Chinese Cuisine, presented the award to Li Qigui.

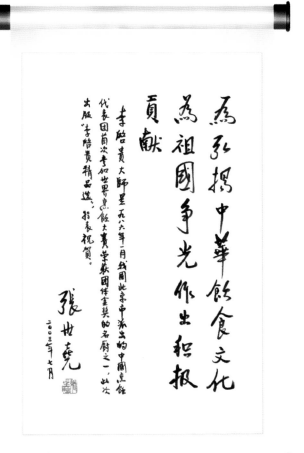

前言

《薪传——中国烹饪名家》系列丛书，是北京烹饪协会在北京市商务局的支持指导下，发起的"中华烹饪技艺传承文化工程"项目的一个重要组成部分。

中国烹饪文化源远流长，博大精深。早在一百多年前，孙中山先生就曾比较中国与西方饮食文化说："中国近代文明进化，事事皆落人之后，唯饮食一道之进步，至今尚为开明各国所不及。中国发明之食物，固大盛于欧美；而中国烹调法之精良，又非欧美所可并驾。"北京是首批国家历史文化名城、中国四大古都之一和世界上拥有世界文化遗产数最多的城市。北京也是中华餐饮文化的荟萃之地，各大菜系的名师大厨辈出，薪火相传不息。

《薪传——中国烹饪名家》系列丛书，所选的烹饪名家都是在北京餐饮行业服务多年，在业界有一定影响，特别是在烹饪技艺传承方面取得卓著成绩的名厨名师。近些年来，国家一直在倡导：各行各业"都要培育精益求精的工匠精神"。倡导工匠精神，离不开传承。工匠精神落在个人层面，就是一种职业精神，它是从业者职业道德、职业能力、职业品质、职业价值取向和行为的综合体现。中华烹饪技艺薪传的一个重要特点，就是讲求师承，注重"手把手""一对一"的言传身教。所有的菜系或老字号名吃名店，都是依靠一辈辈名厨大师带徒弟或家族传承的方式，将其精湛的烹饪技艺、独门绝技，包括诚信敬业，精益求精，不断追求完美和极致的职业精神传承给后继者。这是中华烹饪技艺薪传弥足珍贵的优良传统，也是一切有成就有建树的名厨大师所具备的特质和精神。本套丛书所突出的宗旨和主题，就是对这些宝贵的精神财富和优良传统加以传承和弘扬。

丛书总体规划和主旨统一，每本分册又独具特色。书中将通过各位烹饪大师的从业经历、所在老字号名店的菜系文化、师承谱系，以及与之相关的历史沿革、文化积淀、经典名菜等众多内容和大量精彩纪实照片，突出其德艺双馨、匠心匠艺、独门绝技，并以地方菜系融入北京文化为脉络进行梳理，较为全面地将各位烹饪名家为传承中华烹饪技艺所做出的执着追求、取得的成绩，以及对首都饮食文化、社会生活的贡献和影响，个性鲜明地展现给行业同道和广大读者。同时在经典菜品介绍方面，也以更新的角度为拓展大家对烹饪的认知、完善行业的品位德行等做了有益的尝试。

总之，致力做好中华烹饪技艺传承工作，是北京烹饪协会义不容辞的责任。

Preface

Keep the Fire Going—The Chinese Culinary Masters, a series of books, is an important part of the Cultural Project of Inheriting Chinese Cooking Skills initiated by Beijing Cuisine Association under the support and guidance of Beijing Municipal Commission of Commerce.

The extensive and profound Chinese cuisine culture has a long history. More than one hundred years ago, Mr. Sun Yat-sen compared Chinese and western cooking cultures and said, "The modern civilization of China falls behind in all aspects, except cuisine, which is far beyond enlightened countries. Chinese food is very popular in Europe and America, but they cannot master the superior skills of Chinese cuisine". As one of the first national historical and cultural cities and four ancient capitals of China, Beijing also has the maximum world cultural heritage in the world. In addition, Beijing is the hub of Chinese cuisine cultures with inheritance from a large number of famous chefs in different cuisines.

Famous chefs selected in *Keep the Fire Going—The Chinese Culinary Masters* all work in the catering industry of Beijing for many years with influence in the industry. In particular, they are distinguished in the inheritance of cuisine skills. In recent years, the state has been advocating the excelsior craftsmanship in all walks of life. The initiation of craftsmanship cannot be carried out without inheritance. Craftsmanship is the professional spirit for individuals, which is the overall manifestation of the professional ethics, ability, quality, value orientation and performance of practitioners. One important characteristic in the heritage of Chinese cuisine skills is the succession from teachers to disciples, which emphasizes teaching hand by hand and one to one. In well-known restaurants of different cuisines or time-honored brands, the superior cuisine skills and unique techniques, including the professional spirits of integrity, dedication, excellence and continuous pursuit of perfection and extreme, are carried forward from famous chefs to disciples or families from generation to generation. It is a valuable tradition in the heritage of Chinese cuisine skills. It is also the peculiarity and spirit of all successful famous chefs and masters. The purpose and theme of these books are to inherit and carry forward such valuable spiritual wealth and excellent tradition.

With overall planning and unified theme, these books are characterized by unique features in different fascicules. The professional excellence, moral integrity, ingenuity, craftsmanship and unique skills of famous chefs are highlighted through working experience, cuisine cultures of their time-honored brands, relationship with teachers, relevant history, culture, classic dishes and many documentary images. In addition, the blending of local cuisines into Beijing cultures is surveyed to show their persistent pursuit and achievements in inheriting Chinese cuisine skills and their contributions to and influence on the cooking culture and social life of Beijing to chefs and readers. Classic dishes are described from new viewpoint to expand the knowledge about cuisine and perfect industrial virtues.

In short, it is the obligatory duty of Beijing Cuisine Association to inherit Chinese cuisine skills.

我最想说的话

  我是一名厨师，跟红案白案、煎炒烹炸打了大半辈子交道。

  现在回想起来，您说人生像不像一桌宴席：既有酸甜苦辣咸五味俱全，又分凉菜、热菜、汤菜、甜品各个阶段；既有压桌小菜的随性率真，也有头盘的高光时刻；饮食烹饪的过程，不就是人生磨砺的过程吗？

  真是"三句话不离本行"，您看，我一张嘴，就拿宴席打起了比方。

  其实，"民以食为天"，对于大多数上了年纪的中国人来说，饥饿的记忆是刻骨铭心的。

  旧社会，我的祖父带着一家人从山东逃荒到了北京，就是为了吃口饱饭。1969年1月23日，还不满17岁的我，响应毛主席"上山下乡"的号召，到农村插队参加工作。从那时起至今，已然51个年头，转眼我也到了"发挥余热"的岁数。2020年，在我年满68岁的时候，北京市商务局推出了"薪传"《中国烹饪名家.李启贵：京菜》，令我惊喜交加。这既是对我几十年工作的鼓励，也是对我今后努力为祖国的烹饪事业服务、做好"传帮带"工作的鞭策。我生在新社会，长在红旗下，是党的培养才使我在餐饮服务行业的岗位上，为人民做了自己应该做的工作。是党的具有伟大历史转折意义的十一届三中全会，给了我事业腾飞的机遇：1978年，根据工作需要，我来到新单位正阳春（泰丰楼的前身），亲身感受到改革开放之初的蓬勃朝气。当时在三中全会精神的鼓舞下，北京市饮食服务总公司开始抓技术，举办了北京市第一期山东菜系高级进修班，我有幸成为其中的一名学员。总公司外事处处长兼饮食处副处长李正权，出任我们进修班的班主任。李老在改革开放的进程中，积极推动落实恢复老字号的好政策，京城老字号泰丰楼在正阳春的基础上得以重生。我的烹饪事业也从此有了更为广阔的舞台。

  1986年，我作为新中国派出的第一个烹饪代表团骨干成员，首次在卢森堡举行的奥林匹克第五届世界烹饪大赛上荣获金牌、金牌证书和金牌纪念杯，为祖国和中华烹饪赢得了世界级的荣誉。1993年，我参加了全国第三届烹饪大赛，荣获金牌并荣获"全国优秀厨师"称号。1994年，我参加了川鲁苏粤四大菜系的大师赛，再次荣获金奖第一名。

  荣誉来之不易，饮水尚须思源。回头看看我走过的路，有两个字涌上我的心头："感谢"。我要感谢伟大的中国共产党，感谢改革开放，感谢鼎力支持我的李正权先生，感谢我的启蒙师父艾长荣、抻面的周子杰恩师、助我开办泰丰楼的马德明恩师、教我雕刻的崔技良恩师、蔡启厚恩师，更要感谢亲自指导我多次参加各项大赛并荣获金奖的恩师王义均师父。师父的恩情永远温暖着我的心，我铭记不忘。我还要感谢所有和我工作过的同事们，他们都给了我很多的帮助和支持。真诚感谢曾经帮助、支持、培养我的领导，是你们给了我越来越广阔的人生舞台。未来，我要和烹饪界同行，和我的徒弟们，在以人民为中心的烹饪事业上，启于守正，贵在创新，把菜品做得更好，把人生的宴席办得更精彩。

李启贵

2020年10月1日

What I Want to Say Most

As a cook, I have dealt with cooking dishes, making noodles & pastries, frying, stir-frying, boiling and deep-frying for most of my life.

Looking back now, do you think life is like a banquet for it has sweet, sour, bitter, spicy and salty flavors, and it can also be divided into stages including cold dishes, hot dishes, soups and desserts? There are not only the casual frankness of cold side dishes, but also the highlights of the first dish. Isn't the process of eating and cooking the process of steeling ourselves throughout the life?

One can hardly open his/her mouth without talking shop. So do I. I take banquets as an analogy when starting a talk.

In fact, "food is the first necessity of the people". For most elderly Chinese, hunger is unforgettable.

In the old society, my grandfather and his family fled from Shandong to Beijing just for a living. On January 23, 1969, I still under the age of 17, responded to Chairman Mao's call to "go and work in the countryside and mountain areas" and went to live and work in a countryside. It has been 51 years since then, and in the twinkling of an eye, I have reached the age of contributing my remaining energy. In 2020, when I was 68 years old, the Beijing Municipal Commerce Bureau launched "Keep the Fire Going" series Li Qigui Volume, which pleasantly surprised me. This is not only an encouragement to my decades of work, but also a spur to my future efforts to serve the cooking cause of the motherland and to do a good job of "passing on learnings to the coming generations". I was born in a new society and grew up under the red flag. It was the Party that cultivated me to duly serve the people in my post in the catering service industry. It was the Third Plenary Session of the 11th CPC Central Committee, which was of great historical turning significance, that gave me the opportunity to develop my career successfully. In 1978, as required for the work, I came to Zhengyangchun Restaurant (the predecessor of Taifeng Restaurant), a new unit, and personally felt the vigor and vitality at the beginning of the Reform and Opening Up. At that time, inspired by the spirit of the Third Plenary Session of the CPC Central Committee, Beijing Catering Service Corporation began to stress skills and held the first advanced training class of Shandong cuisine in Beijing. I was lucky enough to be one of the students. Li Zhengquan, Director of the Foreign Affairs Department and Deputy Director of the Catering Department of Beijing Catering Service Corporation, served as the head teacher of our advanced training class. In the process of Reform and Opening Up, Mr. Li actively promoted the implementation of the good policy of restoring time-honored brands. Taifeng Restaurant, a time-honored brand in Beijing, was reborn on the basis of Zhengyangchun Restaurant. My cooking career had also stepped into a broader stage since then.

In 1986, as a key member of the first cooking delegation sent by the People's Republic of China, I won the gold medal, the gold medal certificate and the gold medal commemorative cup for the first time in the 5th IKA held in Luxembourg, winning world-class honor for China and Chinese cooking. In 1993, I took part in the third national cooking competition and won the gold medal and the title of "National Excellent Chef". In 1994, I took part in the master competition of the four major cuisines such as Sichuan, Shandong, Jiangsu and Guangdong cuisines, and won the first gold medal again.

Honors are hard-won, and I'm grateful for honors received. Looking back at the road I walked, I come up with two words, "Thank you". I would like to thank the great Communist Party of China, the Reform and Opening Up, Mr. Li Zhengquan, who fully supported me, Mr. Ai Changrong, who enlightened me at the very beginning, Mr. Zhou Zijie, who gave me helpful instructions on making noodles, Mr. Ma Deming, who helped me in setting up Taifeng Restaurant, and Mr. Cui Jiliang and Mr. Cai Qihou, who taught me food carving. I would particularly appreciate Master Wang Yijun, who personally guided me to participate in various competitions and win gold medals, and his kindness always encourages me and I will never forget it. I would like to extend my thanks to all my colleagues for their help and support. I would sincerely thank the leaders who helped, supported and trained me, providing me with a broader and broader stage of life. In the future, I will keep righteousness and focus on innovation in the people-centered culinary career with peers in the culinary circle and my disciples, so as to present better dishes and a more wonderful banquet of life.

Li Qigui
October 1, 2020

李启贵与恩师王义均。
Li Qigui and his mentor Wang Yijun.

李启贵与恩师艾长荣。
Li Qigui and his mentor Ai Changrong.

李启贵与恩师马德明。
Li Qigui and his mentor Ma Deming.

廣採博取
精益求精

师 王義均
二〇〇三年秋

李启贵与恩师李正权。
Li Qigui and his mentor Li Zhengquan.

李启贵与恩师蔡启厚。
Li Qigui and his mentor Cai Qihou.

李启贵与恩师崔技良。
Li Qigui and his mentor Cui Jiliang.

一九八六年中国烹饪代表团首次参加奥林匹克世界烹饪大赛，李启贵大师获得金牌、金牌纪念杯、金牌证书。

1986年，李启贵大师在卢森堡为首相亲自烹制首相宴。

In 1986, Li Qigui personally cooked for the Prime Minister's banquet in Luxembourg.

2000年，李启贵大师应邀到上海锦江饭店为挪威首相烹制名菜。

In 2000, Li Qigui was invited to Jin Jiang Hotel Shanghai to prepare well-known dishes for the Norwegian Prime Minister.

2008年8月9日，喜迎百年奥运，李启贵大师为德国前总理施罗德烹制中华鼎宴。
On August 9, 2008, Li Qigui cooked a Chinese tripod banquet for the former German Schroeder Chancellor to welcome the centennial Olympics.

2012年，李启贵大师在万寿寺为澳大利亚前总理霍克烹制中华八珍宝鼎专利名菜。
In 2012, Li Qigui cooked the patented Chinese Eight Delicacies Banquet at Wanshou Temple for former Australian Prime Minister Robert James Lee Hawke.

目录 Contents

第一回 Chapter One	厨行新秀 名师门下学绝技 神秘晚宴 四十八年未解开 A Promising Young Cook Learning Unique Skills from Famous Teachers The Mysterious Dinner Unsolved for Forty-eight Years	1
第二回 Chapter Two	得偿夙愿 抻龙须三冬两夏 神秘来信 泰丰楼重振雄风 Realizing the Long-cherished Wish and Pulling the Longxu Noodles for Three Winters and Two Summers A Mysterious Letter Revitalizing Taifeng Restaurant	25
第三回 Chapter Three	雪中送炭 竞技场因谁得救 勇夺首金 开幕式为他推迟 Timely Assistance Rescuing the Competition Winning the First Gold Medal, He Made the Opening Ceremony of the IKA Delayed	43
第四回 Chapter Four	进退有据 重金不卖一招鲜 大师驾到 京派鲁菜迷港岛 Adhering to the Principle and Refusing to Sell Cooking Skills Even at A Huge Sum of Money Beijing-style Shandong Cuisine Attracting Great Attention in Hong Kong	59
第五回 Chapter Five	身教为先 盛宴亲烹三道菜 天伦布阵 六大名厨闹京都 By Setting Examples, He Cooked First Three Dishes for All Important Banquets With Exquisite Layout and Design, Six Famous Chefs Competing in Beijing	77

| 第六回
Chapter Six | 妙用灵芝 绝品迷住四海客
情系宝鼎 八珍融入五千年
Fabulous Dishes with Ganoderma Lucidum Enchanting Tourists across the World
Eight Treasures in Tripod Integrating the History of 5,000 Years | 93 |

| 第七回
Chapter Seven | 遍访名师 红白绝技都在手
感恩传递 无私课徒海胸襟
Visiting Many Famous Masters and Mastering Dish-cooking & Noodle-making Skills
Passing on Unique Skills with Gratitude & Delivering Unselfish Lessons | 109 |

| 第八回
Chapter Eight | 良乡传艺 授徒先从磨刀起
克绍箕裘 弟子烹坛屡夺魁
Passing on Culinary Skills from Liangxiang and Starting from Sharpening Knives
Carrying Forward the Cooking Skills and Empowering Apprentices to Win Medals | 123 |

经典菜品　　　　141
Classic Dishes

年　表　　　　177
Chronology

第一回

Chapter One

厨行新秀　名师门下学绝技
神秘晚宴　四十八年未解开

1982年6月16日，对于李启贵来说，注定是他人生当中和烹饪路上非常重要的一天。那天中午，李启贵从正阳春酒楼下班，上街置办了四样礼物：两瓶白酒、两桶茶叶、一盒稻香村点心和一兜子苹果香蕉，大步流星地赶奔永定门内大街。一路上李启贵抑制不住激动的心情，经李正权先生引荐，他今天要正式拜丰泽园鼎鼎大名的王义均师傅为师，继续深造烹饪技艺。这是他期盼已久的大事情。

曾经在北京市1979年第一届山东菜系培训班深造过的李启贵，在京城的青年厨师中已然崭露头角，小有名气。这次拜王义均为师，学习最正宗的山东菜，李启贵真可谓登堂入室，因而倍感珍惜。

王义均大师当时住在永定门里路东的一个小院里，靠东北角的两间平房。李启贵拜师的时候，是改革开放之初，那时候还没有恢复旧年间的那套习俗：在饭店酒楼"摆枝"，请来八方宾朋见证，有引师、保师、代师，给师父师娘行叩拜大礼……这些与当时的社会风气格格不入。人与人之间的关系既简单又淳朴，徒弟尊师敬师，师父包教包会，提高烹饪技艺，"更好地为人民服务"，这，就齐活了！

王义均推门一看，门外站了个浓眉大眼、身材魁梧的小伙子，一脸喜兴，就知道是李正权处长介绍来的徒弟李启贵。人和人是有"眼缘"的，王义均一眼就喜欢上了这个脑门上冒着汗珠儿的山东小老乡。爷儿俩聊得挺投机。聊到太阳偏西，李启贵起身告辞，老爷子说什么不让走，非要留他吃饭。李启贵说："师父那我做吧。"老爷子说："启贵别看你年轻，你手上没我快。"结果王义均用了七分钟，连凉带热做了11个菜，简直是神速！李启贵在边上看呆了，或许这也是王师傅特意在新徒弟面前露一手。老爷子拿出一瓶高庄的白酒，两个牛眼小酒盅，爷儿俩一口一个，边聊边喝，一斤多酒全喝了。

从这儿之后，丰泽园后院的长廊就成了李启贵的"课堂"。王义均忙完厨房的工作，就到后院的长廊上给李启贵传授烹饪技艺。现在回忆起当时的情景来，李启贵真情流露，说了八个字："师恩浩荡，饮水思源。"

A Promising Young Cook Learning Unique Skills from Famous Teachers
The Mysterious Dinner Unsolved for Forty-eight Years

June 16, 1982, for Li Qigui, was destined to be a very important day in his life and cooking career. At noon that day, Li Qigui went off work from Zhengyangchun Restaurant and went into the street to buy four gifts, two bottles of liquor and spirits, two barrels of tea, a box of Daoxiang Village dim sum and a bag of apples and bananas. He marched to Yongdingmen Inner Street. Along the way, Li Qigui could not restrain his excitement. Recommended by Mr. Li Zhengquan, he would formally acknowledge Mr. Wang Yijun, the famous chef of Fengzeyuan Hotel, as his master that day to further his cooking skills, to which he has been looking forward for a long time.

Li Qigui, who studied in the first Shandong cuisine training class in Beijing in 1979, had already shown up prominently among the young chefs in Beijing as a minor celebrity. By taking Wang Yijun as his master to learn the authentic Shandong cuisine, Li Qigui would become more proficient and felt pretty honored.

Master Wang Yijun was living in a small courtyard in the east of Yongdingmen, two single-storey houses in the northeast corner. It was at the beginning of the Reform and Opening Up that Li Qigui took Wang Yijun as his master. Then, the customs of the old years had been removed, i.e., holding an acknowledgement ceremony in restaurants, inviting guests and friends from all walks of life to witness, having recommender, guarantor and substitute master performing their duties, kowtowing to master and his wife and others, which were out of tune with the social atmosphere at that time. The relationship between people was simple and honest. Apprentices respected their masters while masters taught apprentices to improve their cooking skills and "served the people better". This was all!

Opening the door, Wang Yijun saw a burly young man with heavy eyebrows and big eyes standing outside the door. He knew that this pleasing man was Li Qigui, the apprentice introduced by Director Li Zhengquan. People have eyes-affinity with each other. Wang Yijun took to this young fellow-villager from Shandong with sweat on his forehead at a glance. With a lot of things in common, they talked till the later afternoon, and then Li Qigui stood up to say good-bye. But Master Wang would not let him leave and kept him for dinner. Li Qigui said, "Master, let me cook". Master Wang replied, "Qigui, young as you are, I would be faster than you." As a result, it took Master Wang seven minutes to cook eleven dishes, both cold and hot. It was amazing! Li Qigui was stunned at this. Perhaps this was also Master Wang's special show in front of this new apprentice. Master Wang took out a bottle of Gaozhuang liquor and spirits, and two small liquor cups with cattle eyes. They had one cup for each drink, chatting and drinking, with a half kilogram of liquor and spirits bottomed out.

Since then, the long corridor in the backyard of Fengzeyuan Hotel had become Li Qigui's "classroom". Wang Yijun went to the corridor in the backyard to teach Li Qigui cooking skills after he finished his kitchen work. Recalling the scene at that time, Li Qigui sincerely said, "With mighty kindness from my master, I would never forget him since when one drinks water, one must not forget where it comes from."

为什么这么说呢？李启贵认为，有人说师恩如山，是因为高山巍峨使人崇敬；有人说师恩似海，是因为大海浩瀚让人畅想。在李启贵看来，王义均不仅有炉火纯青的烹饪绝艺，更有倾囊相授、海纳百川的胸怀，还有教学相长、转益多师的格局。几年后，李启贵之所以能作为中国代表团的主力，在第五届世界奥林匹克烹饪大赛上夺冠，与王义均的悉心栽培和无私教诲密不可分。

王义均大师"拿手绝活儿"甚多，他毫不保留地传授给了爱徒李启贵。限于文章篇幅原因，咱们择其要者，重点说三道王大师传授的名菜：炸空心龙虾球、葱烧海参、百花珍珠鱼。

王义均和李启贵确立师徒关系后，如果说丰泽园的后院长廊是口传心授的第一课堂，那么右安门第三旅馆李启贵的周转房里，就是实际操练的教学现场。在这里，王大师手把手教了李启贵整整八十天。

李启贵先跟师父学吊汤。有道是"唱戏的腔，厨师的汤"。中国厨师吊出的汤讲究味道鲜醇而汤清如水。李启贵凭着师父教的一道凤蓉茉莉竹荪清汤，夺得了第3届全国烹饪大赛金牌，并获得了全国优秀厨师称号。这道汤的原料包括水发上好的天然竹荪、水发香菇、青菜叶、胡萝卜、鸡里脊、清汤、茉莉花、湿淀粉、鸡蛋清、精盐、料酒、葱姜油等。具体做法是，竹荪改刀成象眼形，鸡里脊去筋皮，加入蛋清、精盐、料酒、葱姜油，打成蓉，放入羹匙中抹上油，点缀上花和香菇，上屉蒸熟成凤蓉。竹荪用水焯透，入汤锅，加入调味料，开后去沫，盛入汤窝中，放入凤蓉、茉莉花，加盖即可上桌食用。这道菜的特点是，打开盖后花香四溢，汤鲜味美，质地软嫩，入口清爽。王义均告诉李启贵这道菜的关键步骤，就是茉莉花什么时候放：临上桌前的最后10秒钟再放，到桌上一掀汤窝盖儿，茉莉花的清香味顺着热气一下子散发出来，满堂香！放早了便无此神效。李启贵回忆道："当年比赛正值冬季，这几朵珍贵的茉莉花，是他骑车从丰台花乡的地窖子里淘换来的。"

王义均大师最出名的一道菜是"葱烧海参"。这道菜的选料包括水发海参、大葱、姜、蒜、香菜、青蒜、葱姜米、鸡汤、酱油、精盐、淀粉、白糖、料酒、食用油等。具体做法是海参洗净，大个儿的一分为二，用沸汤焯过，再加味儿煨一会儿，大葱切成6厘米长的段儿，用油炸成金黄色，放入碗内，加入鸡汤、料酒、酱油、白糖等，上屉蒸一两分钟，滗去汤汁。油烧热，下入葱姜蒜片和香菜炸成葱油。炒锅上火炒糖色，下入葱姜米、海参、葱油、料酒颠翻均匀，放入鸡汤和调味料煨好后，用淀粉勾芡，加入葱段，撒入青蒜段即成。

So for what? Li Qigui believes that it's said that the kindness received from masters is as great as a mountain, which is because lofty mountains are respectable; it's also said that the kindness received from masters is as deep as oceans, which is because oceans are so vast as to enable people to fall into a reverie. In Li Qigui's view, Wang Yijun not only has perfect cooking skills, but also has a mind of delivering instructions without reservation and keeping tolerance to all differences, as well as a pattern of teaching and learning from each other and turning to learning more from others. A few years later, Li Qigui was able to win the 5th IKA as the main force of the Chinese delegation, which was inseparable from Wang Yijun's meticulous cultivation and selfless teaching.

Master Wang Yijun has many "unique skills" and he has unreservedly taught them to his apprentice Li Qigui. Limited by the length of the article, let's choose the most important one, focusing on three famous dishes taught by Master Wang, Fried Hollow Lobster Balls, Scallion-roasted Sea Cucumber, and Pearl Fish with Flowers.

After Wang Yijun and Li Qigui established the mentoring relationship, if the long corridor in the backyard of Fengzeyuan Hotel was the first classroom for oral teaching that inspired true understanding within, then the turnover room of Li Qigui in Youanmen No. 3 Hotel was the actual teaching site, where Master Wang instructed Li Qigui in person for 80 days.

Li Qigui learned simmering soup from his master first. There is a saying that "the soup by a cook is as important as the tune of singing opera". Chinese chefs devote particular care to freshness and mellow while keeping it as clear as water in making soup. Li Qigui won the gold medal in the 3rd National Cooking Competition and the title of National Excellent Chef with the Celosia Jasmine & Dictyophora Clear Soup taught by his master. The main ingredients of this soup include water-fat natural Dictyophora, water-fat Lentinus edodes, green vegetable leaves, carrots, chicken tenderloin, clear soup, jasmine, wet starch, egg white, refined salt, cooking wine, fried scallion-ginger oil, etc. The specific method is to cut Dictyophora into eye-shaped pieces, remove tendon and skin from chicken tenderloin, add egg white, refined salt, cooking wine and fried scallion-ginger oil, mash and put them into spoons, put oil on them, embellish with flowers and mushrooms, and steam into Celosia in a drawer. Blanch Dictyophora thoroughly with water, put it into a soup pot, add seasonings, remove foam after boiling, add Celosia and jasmine flowers, cover it and then serve on the table. This dish is characterized by the fragrance of flowers, fresh and delicious flavor, soft and tender texture and refreshing function after the lid is opened. Master Wang told Li Qigui that the key step of this dish is when to put the jasmine flowers in it, the last 10 seconds before it is served on the table. The clear fragrance of the jasmine flowers suddenly emits along the hot air across the house while the lid of the soup pot is lifted! If the jasmine flowers are put earlier, this magic effect cannot be achieved. Li Qigui recalled, "The competition was held in winter that year, and he rode his bike to get these precious jasmine flowers from the cellar in Fengtai Flower Town."

Master Wang Yijun's most famous dish is Sea Cucumber with Scallion. The ingredients of this dish include water-fat sea cucumbers, scallion, ginger, garlic, coriander, leeks, diced ginger and scallion, chicken soup, soy sauce, refined salt, starch, white sugar, cooking wine, edible oil, etc. The specific method is to wash the sea cucumbers, divide the big ones into two parts, blanch them with boiling soup, add flavor and simmer for a while, cut the scallions into 6cm long segments, fry them into golden yellow, put them into a bowl, add chicken soup, cooking wine, soy sauce, white sugar, etc., steam them in a drawer for a minute or two, and decant the soup. Heat the oil, add scallion, ginger

王师傅倾囊相授，李启贵得到真传。1983年，泰丰楼重张之际，溥杰先生亲往用餐。品尝了李启贵烹制的"葱烧海参"后，老先生称赞此菜"葱香味浓，海参软烂，可称上品"。得知这是王义均师傅的亲授，更是欢喜非常。兴之所至，还现场书赠墨宝："继往开来、发扬光大"，对李启贵寄予了更多的期许。

王师傅扒海参的做法，李启贵也学到了手。"扒蝴蝶海参"，这道菜的菜名记载为明代，原来指海参直接抹刀片即为蝴蝶片，抹泥子后为形似，不但时常脱落，而且远未逼真。经过王义均大师的不断改进，不断完善，这道"蝴蝶海参"真正做到了形神兼备，色香味俱全。李启贵不仅在很多大赛现场展现过这道菜，而且还潜心钻研，使之更加完美，后来他把这道菜师承来的名菜又传给了他的徒弟。

"清蒸百花珍珠鱼"，顾名思义，是一道蒸菜。李启贵从王师傅那儿学会了蒸菜的技巧，不但凭借此菜获得了奥林匹克第五届世界烹饪大赛的金牌，而且至今也是天伦王朝酒店的看家菜。这道菜的关键在刀口的功夫上，当年是王师傅手把手地教李启贵"瓦垄刀"，刀法一定要精细，距离要一样，深度要一样。另外，此菜原料也很讲究，"珍珠"要取大个儿的鲜贝，中间的"红珍珠"要用大虾的虾黄。一般厨师不知道虾黄蒸的时候要放在哪儿。王义均大师告诉启贵，虾黄要放在生菜叶上蒸，还不能过火，蒸好之后切成段儿，然后镶嵌在鲜贝上。清蒸珍珠鱼的汤，一定要用纯好的上清汤，同时严格控制火候，才能达到鱼肉鲜嫩、汤鲜味美、造型美观的效果。

谈到"师恩浩荡"，李启贵提起了他十分看重的一次比赛：川鲁苏粤四大菜系名厨大赛。参赛的名厨来自广州的白天鹅、上海的锦江、四川的陈麻婆和北京的泰丰楼，参加这次大赛的厨师都是参加过世界烹饪大赛并且获得过金奖的选手。李启贵当时做的参赛作品是王义均大师亲传的芫爆墨鱼花，获得了金奖第一名。李启贵先生给我看过一张照片，王义均大师站在一号灶展示厨艺，李启贵在旁边随侍。他解释说，这次大赛他拿了大赛金奖第一名，根据大赛要求，第一名获奖者的指导老师赛后在一号灶展示烹饪技艺，以此类推。

and garlic slices and coriander, and fry into scallion oil. Stir-fry sugar in the frying pan, add diced ginger and scallion, sea cucumbers, scallion oil and cooking wine, turn over evenly, add chicken soup and seasonings, simmer, thicken with starch, add scallion segments, and then put some leeks and garlic segments.

Master Wang instructed Li Qigui without reservation and the latter got the authentic skills from him. In 1983, when Taifeng Restaurant was reopened, Mr. Pu Jie went to dinner in person. After tasting the Sea Cucumber with Scallion cooked by Li Qigui, he praised the dish with "scallion fragrance and soft and tender sea cucumber, which can be ranked as top grade" . He was even more gratifying to learn that this was instructed by Master Wang Yijun. As pleased as he was, he presented his writing on the spot, "Carry forward the past traditions and forge ahead into the future", placing more expectations on Li Qigui.

Li Qigui also learned the method that Master Wang braises sea cucumbers. Braised Sea Cucumbers with Butterfly Sauce was recorded in the Ming Dynasty. It originally meant that the sea cucumbers were directly cut into butterfly oblique slices with sauce, which were similar to butterfly in shape. They not only often fell off, but also were far from realistic. Through the continuous improvement by Master Wang Yijun, this Braised Sea Cucumbers with Butterfly Sauce had truly achieved the unity of form and spirit, color, aroma and taste. Li Qigui not only showed this dish at many competition sites, but also studied it with great concentration to make it more perfect. Later, he passed on the famous dish inherited from his master to his apprentices.

"Steamed Pearl Fish with Flowers", as its name implies, is a steamed dish. Li Qigui learned the skill of steaming dishes from Master Wang. He not only won the gold medal in the 5th IKA with this dish, but also made this dish a specialty of Sunworld Dynasty Hotel. The key to this dish lies in the skills in cutting. In those days, Master Wang taught Li Qigui the cutting in rows of tiles on a roof hand in hand, naming "Walong Cutting". The cutting must be fine, with the same distance and the same depth. In addition, the main ingredients of this dish are also very exquisite. The "pearl" should be taken from big fresh shellfish, and the "red pearl" in the middle should be taken from the shrimp yellow of prawns. The ordinary chef does not know where to put the shrimp yellow when steaming this dish. Master Wang Yijun told Qigui that shrimp yellow should be steamed on lettuce leaves without overheating. After steaming, shrimp yellow should be cut into segments and then inlaid on fresh shellfish. Steamed pearl fish soup must be made of pure and clear soup, and the temperature must be strictly controlled at the same time, so as to achieve the effects of fresh and tender fish, delicious soup and beautiful appearance.

When it comes to the mighty kindness of his master, Li Qigui mentioned a competition that he attached great importance to, the competition of famous chefs of the four major cuisines in Sichuan, Shandong, Jiangsu and Guangdong. The famous chefs participating in the competition came from White Swan Hotel in Guangzhou, Jinjiang Hotel in Shanghai, Chen Mapo in Sichuan and Taifeng Restaurant in Beijing. The chefs participating in the competition were all contestants who participated in the IKA and won gold medals. At that time, Li Qigui's entry was the Stir-fried Cuttlefish Flower with Coriander learnt from Master Wang Yijun, and won the first gold medal. Mr. Li Qigui showed me a photo in which Master Wang Yijun was standing in the No. 1 cookstove to show his cooking skills, with Li Qigui beside him. He explained that he won the first gold medal in the competition. According to the requirements of the competition, the instructor of the first winner showed his cooking skills in the No. 1 cookstove after the competition, and so on.

"爆"菜一般分为汤爆、芫爆、酱爆、油爆四大爆。王义均曾教李启贵做芫爆鸡丝、芫爆墨鱼花、芫爆里脊、油爆双脆、汤爆猪肚、汤爆双脆……"爆"菜同样讲究精湛的刀工，比如猪肚、墨鱼，用刀的深度是肉厚度的三分之二，宽度上刀不能太密，王义均告诉他，有人切得花刀太密，肉一吃水就收缩，这个菜就老了，就垫牙了。油爆猪肚、油爆双脆都讲究"汁爆主料，食后盘内无余汁"，根据原料的特点，外焦里嫩。李启贵学会这套技艺并掌握之后，厨艺大增。一次，房山有位烹饪协会的负责人去泰丰楼见李启贵，点了这道油爆双脆，吃了将近一个小时，临走时看盘里还有几块双脆，他说，"我太爱吃您这个菜了，我不怕您说我什么，我还想把这两块吃了。"说着夹起来一尝，很是吃惊，当时腾地站起来说："唉哟，这菜都这么长时间了，依然是脆嫩适口，真是太棒了！"李启贵当时听了自然心里美滋滋的，过后一想，这都得益于师父王义均的无私传授啊！

提到"酥炸"技艺，也是触类旁通，学会"酥炸"，可以酥炸鱼条、酥炸金糕、酥炸龙尾……酥炸各式菜品。王义均大师告诉李启贵，酥炸的关键在于'糊'的制作，"冬天要把它看好了，夏天不能让它再发了。你要用油控制它的酥，你要用碱控制它的酸。酥的程度要求提起来，掉到案子上'啪'地碎了。而且"糊"要挂匀了，不能有空心儿的，不能一炸在油里打滚儿。"这些看似简单的技巧，没有师父捅开这层窗户纸，可能多少年自己也悟不透。李启贵把师父的亲授绝艺珍藏在记忆里，在很多展示厨艺的场合熟练运用，发挥了很重要的作用。

"从1982到1992年，是我跟师父学习的黄金的十年。从1992到2002年，是在师父的指引下走向辉煌的十年。在这20年当中，我跟师父学的东西太多了、太多了。"李启贵如是说。后来李启贵在卢森堡、日本、法国、西班牙、荷兰、新加坡等国，或是参加大赛，或是展示厨艺，把王义均师父传授给他的这些烹饪方法都演绎得淋漓尽致。

在李启贵的记忆中，第五届世界奥林匹克烹饪大赛、苏鲁川粤名厨大赛和第三届全国烹饪大赛，他夺得了"三连冠"，是他人生的高光时刻。可是，每每站在领奖台上、聚光灯前，他的脑海里总是回想起1982年的那个初夏，回想起师父在周转房的石棉瓦棚子里教他做葱烧海参，回想起去日本工作临行前的头天晚上，在师父家待到夜里两点半，师父给了他大量的资料，像龙苑菜单、山珍海味、鸡鸭鱼肉各种菜的制作方法，熊掌的发制方法，鱼翅的发制方法，

Quick-fried dishes are generally divided into such four major types as quick-fried dishes with soup, quick-fried dishes with coriander, quick-fried dishes with sauce and quick-fried dishes with oil. Wang Yijun once taught Li Qigui to make quick-fried chicken shreds with coriander, quick-fried cuttlefish flowers with coriander, quick-fried tenderloin with coriander, quick-fried double crisp with oil, quick-fried pork belly with soup, quick-fried double crisp with soup …Quick-fried dishes also focus on exquisite cutting skills. For example, for pig belly and cuttlefish, the cutting depth is 2/3 of the thickness of the meat, and the cutting width cannot be too dense. Wang Yijun told him that when someone cuts too densely, the meat shrinks as soon as it absorbs water, and then the dish gets hard-eaten. Quick-fried double crisp with oil and quick-fried pork belly with oil both pay attention to "quick-fry the main ingredient with oil, and there is no juice left in the plate after eating". According to the characteristics of the main ingredients, it is tender with a crispy crust. After Li Qigui learned this skill, his cooking skills greatly improved. Once, a person in charge of Fangshan Cooking Association went to Taifeng Restaurant to meet Li Qigui. He ordered this quick-fried double crisp with oil and ate it for nearly an hour. When he left, he saw that there were still several pieces of double crisp on the plate. He said, he loved this dish too much and wanted to eat these two pieces regardless of others' opinion on him. He was very surprised when he picked it up and tasted it. At that time, he stood up suddenly and said, "Alas, this dish is still crisp, tender and palatable after serving such a long time. It's really great!" At that time, Li Qigui was very happy of course when he heard of that. After thinking about it, he benefited from Master Wang Yijun's selfless instruction.

When it comes to the skill of "crisp frying", it is similar. In case of "crisp frying", you can crisp fry fish sticks, gold cakes, lobster tails... Master Wang Yijun told Li Qigui that the key to crisp frying lies in the production of "paste". "Keep it warm in winter and keep it cool in summer to avoid fermentation. You have to use oil to control its crisp, and use alkali to control its acid. It's essential to make it crisper, so that it falls to pieces on the kneading board. Moreover, the "paste" should be put on evenly without hollow to prevent rolling in oil." These seemingly simple skills may not have been fully understood within many years without the instruction of the master. Li Qigui treasured Master's unique skills in his memory and displayed his cooking skills in many occasions which played a very important role.

"From 1982 to 1992, it was a golden decade for me to learn from master. From 1992 to 2002, it was a decade towards brilliance under the guidance of master. In these two decades, I have learned much from master.", Li Qigui said. Later, Li Qigui took part in competitions or showed his cooking skills in Luxembourg, Japan, France, Spain, the Netherlands, Singapore and other countries, fully demonstrating the Steps taught him by Master Wang Yijun.

In Li Qigui's memory, he won the "three consecutive championships" in the 5th IKA, the Jiangsu, Shandong, Sichuan and Guangdong Famous Chef Contest and the 3rd National Cooking Contest, which were the highlights of his life. However, every time when he stood on the podium and in front of the spotlight, his mind always recalled the early summer of 1982, when Master taught him how to cook sea cucumbers with scallion in the asbestos shingle shed in the turnover room. Recalling the night before he left for work in Japan, he stayed at Master's house until 2: 30p.m., and Master gave him a lot of information, like Longyuan menu, making methods of table delicacies from land and sea, chicken, duck, fish and various dishes, water-fatting of bear paws and shark fins, crab slaughter ... all that was necessary, as if he was afraid that Li Qigui would encounter difficult problems that couldn't be solved when he was alone... "Mighty kindness of Master should not be forgotten", which occurred frequently in Li Qigui's mind.

螃蟹怎么宰杀……可谓应有尽有，仿佛生怕李启贵独自在外会遇到解不开的难题……"师恩浩荡，饮水思源"八个字，在李启贵的脑海中循环往复。

当然，干上厨师这行，其实还有更多的历史记忆。

济南府东平县的庄稼地里，高粱、玉米棒子还没长成青纱帐，就跟晒干的大葱一样死在地里。暴土扬长的官道上，背井离乡的难民迤逦而行，背着老人的，抱着孩子的，挑着扁担箩筐的，一瘸一拐挂着棍子的，十步一回头，外出求生。人流中走着个推独轮车的大个子，看上去三十郎当岁，脸上白白净净，宽脑门儿，浓眉大眼，穿着一身半旧不新的青布裤褂。大个子的独轮车上，一边放着行李铺盖，另一边坐着个小男孩儿。旁边还有个大点儿的男孩跟着大人的脚步走。大个子一脸愁容，抬头看看毒辣辣的日头，叹了口气，天边一丝云彩也没有。低头看了看，身上的青布裤褂出汗出的，湿了好几大片，都是一圈一圈儿的饸饹印。大个子姓李，名叫李守田，身边带着两个尚未成年的儿子。爷儿仨逃荒，是打算投靠北京大兴县亦庄的远房亲戚家求活路。亲戚在亦庄给一家姓韩的地主种着187亩地，到了那儿，就有饭吃了。

凭着这个念想儿，李氏父子靠给人家干活儿打短工混口饭吃，饥一顿饱一顿，风一程雨一程，好几个月才挨到了亲戚家。从此在京南亦庄土里刨食，种地为生，一家人总算安顿下来。后来，小儿子李春秀还被国民党抓去当了几年火头军，在炊事班炒菜蒸馒头蒸窝头，直到新中国成立前夕又跑回来娶亲务农。

1952年8月27日，老李家添丁进口，李春秀得了个大胖小子，一家子可高兴坏了！看着孩子生得方面大耳，结结实实，一脸富贵气象，爷爷高兴地说："这是咱们家在北京出生的第一代啊，希望他以后过上富贵荣华的好日子，就叫启贵吧！"家里人或许没有料到，这个襁褓中睡得小脸通红的李启贵，后来成了享誉世界烹坛的中国十大名厨之一，应验了祖父当年的美好愿望。

1959年，李启贵开始上小学，在亦庄小学读书，就是现在的开发区亦庄小学。1965年小学毕业，在鹿圈中学念初中。第二年暑假就赶上了"文革"爆发，凑凑合合把初中念完，1968年，16岁的李启贵响应伟大领袖毛主席的号召，"到农村去、到边疆去，到祖国最需要的地方去"，作为知识青年上山下乡的一员，被分配到北京市大兴县红星人民公社亦庄大队双桥北队插队。一共干了三年零七个月。在那儿接受了贫下中农再教育，经受了锻炼，感受到了农村广大农民的朴实和艰辛，确实受教育匪浅。

Of course, as a chef, there are actually more historical memories.

In the twenty-seventh year of the Republic of China, there was a great drought in Shandong. The Yellow River stopped flowing. The soil was too dried with cracks and even Spouting Spring in Jinan City dried up. Food is the most important thing for people, however, people at that time had few food at their houses. In addition, Qingdao was just captured by the Japanese army, Shandong was full of wars, escaping from famine with children was the only way for villagers to live.

In the fields of Dongping County, Jinan Prefecture, sorghum and corn cobs were not yet grown up as green gauze tents and died in the fields like dried scallions. On the government-financed road flying soil, the refugees who had left their homes were meandering, some carrying the elderly, some carrying children and others carrying bamboo baskets, hobbling with a stick. They were reluctant to part from their home to survive. Among the crowd, there was a tall man pushing a unicycle. He looked 30 years old, with a clean face, a wide forehead and heavy eyebrows, in a semi-old green cloth trousers and jacket. On his unicycle, there was a little boy sitting on one side with luggage and bedding on the other side. And also a little older boy followed the footsteps of adults. The tall man looked sad, looked up at the scorching sun and sighed. There was not a cloud in the sky. Looking down, he saw his green cloth trousers and jacket sweating wet for several large areas, all of which were printed in circles. The tall man was Li Shoutian and had two underage sons with him. The three men fled from the famine, intending to take refuge with a distant relative in Yizhuang, Daxing County, Beijing, to find a way to live. The relative planted 187 mu land for a landlord surnamed Han in Yizhuang. They would have food for their arrival.

With this in mind, Li and his son would like to work for the landlord as a seasonal laborer to make a living. They had a hungry trip with wind and rain, and it took them several months to get to the relative' house. From then on, they finally settled down for farming in Yizhuang in the south of Beijing. Later, the younger son Li Chunxiu was captured by the Kuomintang as a gang leader for several years, cooking steamed buns in the cookhouse squad. It was not until the eve of liberation that he came back to marry and farm.

On August 27, 1952, Li's family had a new member and Li Chunxiu had a lovely baby. The family was very happy! Looking at the child's solid and decent appearance with a rich face, Grandpa said happily, "This is the first generation of our family born in Beijing. I hope he will live a good life of wealth and splendor in the future. Let's call it Qigui!" The family may not have expected that Li Qigui, who slept red in his swaddling clothes, later became one of China's top ten famous chefs and well-known in the world's cooking circle, fulfilling his grandfather's good wishes in those days.

In 1959, Li Qigui began to attend primary school and studied in Yizhuang Primary School, which is now Yizhuang Primary School in the Development Zone. He graduated from primary school in 1965 and attended junior high school in Luquan Middle School. The summer vacation of the following year was the outbreak of the Cultural Revolution and he managed to finish junior high school. In 1968, 16-year-old Li Qigui responded to the call of the great leader Chairman Mao to "go to the countryside, to the frontier, and to the place where the motherland needs most". As a member of the educated youth who went to the countryside, he was assigned to the Shuangqiao North Team, Yizhuang, Hongxing People's Commune in Daxing County, Beijing. He worked there for three years and seven months. He therefrom received the re-education of the poor and lower-middle peasants, experienced the training, felt the simplicity and hardships of the vast number of rural peasants, and really received a lot of education.

1972年7月，北京市各行业开始招工，"文革"后期包括建筑业、服务业、餐饮业在内的各行各业，都处在严重缺员的状态。李启贵作为第一批招工对象，在填写志愿时选报了餐饮行业。之所以选择干这行，其实是和家族记忆有直接关系的。在李启贵祖父的记忆里，一家人是饿着肚子逃荒出来的，饥饿的烙印无法抹去。在李春秀的记忆里同样如此，同时做炊事员的经历也让他意识到，什么时候也不会饿着做饭的。李春秀认为，厨师是门手艺，而且是个有传承的手艺，一辈子都能干。三百六十行，行行出状元，只要好好干，一样有出息。李春秀夫妻俩思前想后，拿定了主意，就让儿子学厨师。

报名之后，李启贵参加了餐饮业的学习班。学习班的地点就在珠市口西侧路南的板章路。当时组织学习班的学员们参观、实习了人民餐厅、华北楼饭庄等很多酒楼、饭庄，前后将近一个月。1972年7月底学习班结束后，就宣布分配工作岗位。8月初，年方二十多岁的李启贵来到宣武区华北楼饭庄报到。华北楼饭庄的位置就在前门箭楼的西南角，西河沿儿口上就是。

华北楼饭庄在老北京人的记忆里，名气很大。饭庄斜对面就是前门火车站，赶火车、下火车的旅客熙熙攘攘，摩肩接踵。在北京火车站没建成之前，这里是北京城人流最密集的地方。华北楼紧挨着前门大街和天安门广场，自古就是数一数二的繁华地。所以有人说，来华北楼吃饭的客人"带腿的不吃板凳，带翅膀的不吃飞机"，给什么吃什么。华北楼饭庄搞的是日夜经营，楼上有几十间包房，下边的大厅也可以容纳几百人同时就餐。这里每天的销售额，当时在北京市餐饮业是最多的一家。

新中国成立前，华北楼饭庄的老板叫李文元。华北楼的老同事赵杰曾经告诉李启贵，李文元早先在老北京的金星面粉厂工作。这个金星面粉厂是谁家开的呢？是"红色资本家"孙孚凌他们家开的。这个面粉厂在当时很有名，李文元在那儿上班。他每天下班的时候偷偷"顺"点儿面。怎么"顺"呢？他缝一个细长口袋围在腰上，日积月累，积少成多，当时的白面也是贵物，一袋洋面值两块大洋。李文元来了个"长流水不断线"，把这些面粉随"顺"随卖，就此攒下一笔钱。再后来，他就用这笔钱从天津买了一卡车啤酒，转手再倒腾出去，大赚了一笔，开了这座华北楼饭馆。李文元做生意比较精明，号称"小诸葛"。饭庄的成本得多少？利润有多少？能赚多少钱？李文元肚子里的小算盘打得精准无误噼啪直响。1956年，公私合营以后，华北楼饭庄的经理还是他。到了1972年，李启贵参加工作的时候，华北楼饭庄的党支部书记兼总经理已经

In July 1972, various trades and industries in Beijing began to hire workers. In the late period of the Cultural Revolution, all trades and industries, including the construction, service and catering industries, were in a state of serious shortage of workers. Li Qigui, as the first batch of recruiters, chose the catering as his ideal work. The reason why he chose to do this is actually directly related to his family memory. In the memory of Li Qigui's grandfather, the family fled from the famine hungry, and the imprint of hunger cannot be erased. The same is true in Li Chunxiu's memory. At the same time, his experience as a cook also made him realize that cooks would never be hungry. Li Chunxiu believes that chefs are a craft, a craft with inheritance and he can work all his life. Every profession produces its own topmost master. As long as you do a good job, you would have a promising prospect. After making a great deliberation, Li Chunxiu and his wife made up their mind and let their son learn to cook.

After signing up, Li Qigui took part in a class in the catering industry. The place of the class was Banzhang Road, south of Zhushikou West Road. At that time, the students in the class visited and took internship in many restaurants such as People's Restaurant and North China Restaurant for nearly a month. After the class ended at the end of July 1972, the positions of job assignment were announced. In early August, Li Qigui, who was in his twenties, reported to North China Restaurant in Xuanwu District. North China Restaurant was located in the southwest corner of Archery Tower in Qianmen, at the mouth of Xihe River.

North China Restaurant is very famous in the memory of the old Beijingers. Across the street from the restaurant is Qianmen Railway Station. Passengers catching and getting off the train were bustling and crowded. Before the Beijing Railway Station was completed, this was the most crowded place in Beijing. North China Restaurant was next to Qianmen Street and Tiananmen Square and has been one of the most prosperous places since ancient times. Therefore, it's said that guests who visited the North China Restaurant would eat whatever you gave. North China Restaurant operated from day to night. There were dozens of private rooms upstairs and the hall below could accommodate hundreds of people at the same time. The daily sales here was the largest in the catering industry in Beijing at that time.

Before the liberation in 1949, the North China Restaurant was owned by Li Wenyuan, who earlier worked in Jinxing Flour Mill in old Beijing, according to Zhaojie, an old colleague there. Who ran this Jinxing Flour Mill? It was run by Sun Fuling, a "red capitalist". This flour mill was very famous at that time. Li Wenyuan worked there. Every day when he left work, he secretly took some noodles home. How to "take"? He sewed a slender pocket around his waist, and then accumulated a lot over time. At that time, white flour was also expensive and a bag of flour was worth two pieces of flat silver. Li Wenyuan took flour as a routine without stopping and then sold the flour in a process, thus saving a sum of money. Later, he used the money to buy a truck of beer from Tianjin, and made profits by reselling them. He made a lot of money and opened the North China Restaurant. Li Wenyuan was shrewd in doing business and was known as "Little Zhuge". How much did the restaurant cost? How much was the profit? How much money could you make? Li Wenyuan knew it accurately in his mind. In 1956, after the public-private partnership was launched, he was still the manager of the North China Restaurant. By 1972, when Li Qigui took part in the work, the Party secretary and general manager of North China Restaurant had been replaced by Jia Fengting. Manager Jia was in his 30s and 40s at

换成了贾凤亭。贾经理当时三四十岁，正是干事业的好时候。此时的华北楼饭庄已经成为一个包括华北楼、祥瑞饭馆、褡裢火烧店、天津包子、海花园面馆、同益馆等很多店面在内的集团式餐饮企业。

20世纪七十年代初，北京的餐饮业刚刚复苏，由于厨艺人才青黄不接，一些上了年纪的名厨仍然在灶上唱大轴挂头牌。就拿华北楼饭庄来说，后厨的"明星阵容"可谓集一时之盛：厨艺大师艾长荣亲自担任华北楼的店面经理兼厨师长，京城"八大楼"之一老致美楼的"二灶"张发祥和傅作义将军曾经的家厨郭维洲相伴左右。"头砧"于文斌老师傅更是北京城响当当的"砧板第一刀"，备料讲究丝儿细、片儿薄、丁儿匀。于师傅最厉害的绝技就是"拉鸡丝"。拉鸡丝的刀法叫"推拉刀"。于师傅拉鸡丝的时候是在绸子布上切，一是鸡丝切出来干净，二是显得出与众不同的手艺，颇有仪式感。他把这一尺见方的绸子蘸上水，铺到菜墩子上抹平了，把片好的鸡片一片一片地摆在上面，整整齐齐。手上这把使了几十年的菜刀，雪刃飞薄，在绸子和鸡片之间前后推拉切，那是真正的游刃有余。末了，把切好的鸡丝用刀横着一刮，盛在碗里，该上浆的给灶上拿过去。拉完鸡丝，于师傅把绸子啪地一抖搂，用水一投，放好叠起来了，下次切接着用。令人惊讶的是，老爷子不是厨艺表演的时候这样，平时人家备料做生意就是这么干，是常态！当然，于师傅也有脾气，有时候他把鸡丝切好后，手底下没新活儿了，就右手端着碗把鸡丝送到灶上，左手还拿着杠刀板。眼看着花生油热了，一碗鸡丝下去，厨师没划开，鸡丝滚成肉蛋了。他就"啪"地一杠刀板下去，抽厨师傅一家伙，然后骂骂咧咧地回去铺开绸子接茬拉鸡丝去。所以掌灶的师傅们都怕他，毁了他的手艺，他真敢拍你。

鸟随鸾凤飞腾远，人伴贤良品自高。李启贵刚一入行，就在后厨遇到这么多顶尖高手，再加上他聪敏好学，肯下苦功夫，能不进步神速嘛。这就是所谓站在巨人的肩膀上，起点就高。

在华北楼学了一段时间，李启贵就到了门框胡同的祥瑞饭馆，就是后来的瑞宾楼，铺面很大，上下两层楼，厨房里有四个灶，足有两千平方米左右。祥瑞饭馆的褡裢火烧最有名，称得上家喻户晓。据李启贵回忆，听老一辈厨师们说，最早北京市卖褡裢火烧的只此一家，老板叫袁振华。此人和他哥哥过去是在廊坊头条开金店的。1956年公私合营，金店不让开了。哥儿俩一琢磨，干点儿嘛呢？就在廊坊二条烙这褡裢火烧卖。王植田的褡裢火烧就是从那儿学起来的。王植田是袁振华的伙计。

that time, which was a good time to do business. From then, the North China Restaurant had become a group catering enterprise including North China Restaurant, Xiangrui Restaurant, Dalian Baked Cake Restaurant, Tianjin Steamed Bun, Haihuayuan Noodle Restaurant, Tongyi Restaurant and many other stores.

In the early 1970s, Beijing's catering industry had just recovered. Due to the shortage of culinary talents, some elderly and famous chefs still played the leading role in the industry. Taking North China Restaurant as an example, the "star lineup" of the coming chefs could be described as a peak of the time, Cooking master Ai Changrong personally served as the manager and chef of North China Restaurant, with assistance of Zhang Faxiang, secondary chef of the former Zhimei Restaurant, one of the "eight restaurants" in Beijing, and Guo Weizhou, the former chef of General Fu Zuoyi. Old Master Yu Wenbin was the famous "First of Chopping Board" in Beijing City. The materials should be fine in shreds, thin in slices and even in dices. Master Yu's best stunt was "cutting chicken shreds". The method of cutting chicken shreds is called "pushing and pulling". When Master Yu pulled the chicken shreds, he cut them on the silk cloth, for which the chicken shreds were cut clean as well as unique in craftsmanship with a sense of ceremony. He dipped the one-foot square silk in water and spread it on the chopping board to smooth it out. He placed the prepared chicken slices on it one by one, neatly. This kitchen knife in hand, which had been used for decades, had a thin sharp blade and pushed and cut back and forth between silk and chicken slices. It was really adept. In the end, scrape the cut chicken shreds horizontally with a knife and put them in a bowl. Take the starched chicken shreds to the cookstove. After pulling the chicken shreds, Master Yu shook the shreds with a snap, put them in water, folded them up, and used it next time. Surprisingly, the old man performed in this way in ordinary work instead of only giving a demonstration. Of course, Master Yu also has a temper. Sometimes after he had cut the chicken shreds, there was no other job at hand, so he carried the bowl in his right hand to the cookstove with the chopping board in his left hand. Watching peanut oil was heated, the cook put a bowl of shredded chicken in it and failed to make it evenly, so that the shredded chicken rolled into meat eggs. Master Yu would pat the cook with the chopping board and then critically went back to spread the silk and pull the chicken shreds. Therefore, the chefs were all afraid of him since he would really beat you in case of destroying his craft.

A bird can fly high with the phoenix, one will be edified being with the virtuous. As soon as Li Qigui started his career, he met so many top experts in the kitchen. In addition, he was intelligent and eager to learn and willing to work hard so he made rapid progress. This was the so-called standing on the shoulders of giants, with a higher starting point.

After studying in North China Restaurant for a period of time, Li Qigui arrived at Xiangrui Restaurant in Menkuang Lane, which was later Ruibin Restaurant. The Restaurant was very large, with two floors up and down. There were four cookstoves in the kitchen, covering about 2,000 square meters. Xiangrui Restaurant was the most famous for its Dalian baked cakes and was widely known. According to Li Qigui, the older generation of chefs said that this restaurant was the first one in Beijing to sell Dalian baked cakes. It's owned by Yuan Zhenhua. This man and his brother used to open a gold store in Lane I, Langfang. In 1956, the public-private partnership prevented the gold store from operating. The two brothers thought about a business. Then they made and sold baked cakes in Lane II, Langfang. This is where Wang Zhitian learned the skill, who was Yuan Zhenhua's buddy.

李启贵来祥瑞的时候，经理是王正英。副经理是艾长荣和刘保成。祥瑞原来的裎裢灶上有罗瑞林、陈炳湘、刘满玉三位师傅，王植田最年轻。王植田当时住在大栅栏大清风巷，他的姐夫李德才是全聚德赫赫有名的面点师。之所以让艾长荣带着李启贵等年轻人调到祥瑞，是因为祥瑞当时正在装修，准备重张，缺少炒菜的厨师。

李启贵调到祥瑞的时候，饭馆的装修进入尾声，还没正式开业。李启贵就跟面案组组长王植田在裎裢火烧这边工作，负责包包子。天津包子一两仨，一个包子24个褶儿，每两三个面剂儿是一两五。蒸包子看屉的同时，还要会使吹风机烧烟煤，在蒸锅上端屉，一次放4屉。今天你负责上屉，就是要看屉、刷锅、投屉布，刷笼屉，然后把火拢着，等锅开了上屉、码屉、下屉，叫服务员往出捡包子这一整套活计。就是说哪天你负责轮班儿，今天你负责使碱，就是用大钢盆揭面。头一天就要把这面里放上温水，然后把老肥面抄起来，和开，然后把新面粉放到里边，不是往下搋，而是往上抄，双臂"十字"形来回抄面。抄完了之后再搋，然后再把盖钢盆的布投湿了，上面一蒙，醒着。留着第二天用。这个活儿，既是力气活，又是技术活。第二天上午十点上班，来了就在面里使碱。使碱讲究"一拍二看三听"。什么叫一拍？一拍就是拍拍面团，听听是不是声音发脆，发脆就说明碱使小了。二看就是烫碱，把铲子往煤火上一烧，把这面这儿一揪、那儿一揪，放到一块一搓，在热铲子上一烫，就知道碱使得均不均。三听，碱使合适了，一拍，你一听面团"砰砰"的声音，就是面"说话"了：可以蒸包子啦！

使好碱的面团，拿刀一条一条地切出来，然后手揪剂儿，啪啪啪！揪成鸡蛋大小的块儿。两个人擀皮儿，十来个人包。天津包子很讲究，手法各不相同，这让年轻的李启贵大开眼界。有的师傅是托在手心里包，有的师傅是让这包子在5个手指尖儿上转。经理王正英包出的24个褶儿，跟刀裁的一样，从中间一刀剁下去之后，包子的嘴儿和底儿薄厚一样。擀皮的师傅三下就出一个皮儿，这么多人包，皮儿完全供得上用，手艺确实高超。

1972年11月底，祥瑞正式开业，李启贵离开白案，跟着艾长荣师傅上灶炒菜了。李启贵上班后一直住在门框胡同24号的大宿舍，离单位很近，所以每天他都早来晚走。早上六点多起来，先给砧板王有为师傅倒尿盆，然后到祥瑞上班，先给师傅们掏炉坑、劈劈柴、拢火。头灶是师傅艾长荣，二灶就是李启贵。李启贵早上把这一套活都弄好了之后，把汤锅烧开了，鸡鸭肘子煮上吊汤。给师傅

When Li Qigui came to Xiangrui Restaurant, Wang Zhengying was manager and Ai Changrong and Liu Baocheng were deputy managers. Xiangrui Restaurant originally had three masters, Luo Ruilin, Chen Bingxiang and Liu Manyu, and Wang Zhitian was the youngest. Wang Zhitian lived in Daqingfeng Lane, Dashilan, and his brother-in-law Li Decai was Quanjude Restaurant's famous pastry chef. Ai Changrong and other young people such as Li Qigui were transferred to Xiangrui Restaurant since Xiangrui Restaurant was decorated and prepared for re-opening, and there was a lack of cooks.

When Li Qigui was transferred to Xiangrui Restaurant, the decoration of the restaurant was coming to an end and had not yet officially opened. Li Qigui worked with Wang Zhitian, head of the noodle team for Dalian baked cakes, being responsible for wrapping steamed buns. Three Tianjin steamed stuffed buns was 50g, one steamed stuffed bun has 24 folds, and three doughs for 50g Tianjin steamed stuffed buns were 150g weigh. While steaming stuffed buns, you should also make the blowing machine burn bituminous coal, and put 4 drawers at a time in the upper drawer of the steamer. In other words, if you were on duty, you should be responsible for a whole process, i.e., watching the drawer, brushing the pan, washing the cloth in the drawer, brushing the steamer, and gathering the fire; then loading, arranging and unloading steamed stuffed buns and asking the waiter to pick up the steamed buns. That's to say, if you were in charge of shifts, you should in charge of making alkali, that's, using a large steel basin to make dough. In the previous day, you should put warm water into flour, then mix yeast evenly with the new flour in a bottom-up and crossed manner. After mixing, you should hold it again, then wet the cloth covering the steel basin and cover it on the top for fermentation till the next day. This job was both physical and technical. At 10 o'clock in the next morning, he came to the restaurant and added alkali in the dough. Adding alkali followed the rule of "patting, watching and listening". What is patting at first? It's to pat the dough to see if it sounds brittle, which means the alkali is less. Second, watching is to scald alkali. Burn the shovel on the coal fire, take small doughs from different parts, and rub them together. When you scald the dough the hot shovel, you would find whether the alkali was uneven. Third, listening means you can pat the dough to see whether the alkali was appropriate. If you listen to the dough with "bang bang" sound, it's ok, and you can steam the stuffed buns!

Make the dough with appropriate alkali, cut into pieces one by one with a knife, and then divide into small parts with your hand! Divide it into egg-sized pieces. Two people rolled pieces and a dozen people wrapped. Tianjin steamed stuffed buns were very exquisite and had different techniques, which had broadened young Li Qigui's horizon. Some chefs hold the stuffed buns in their palms, while others let the stuffed buns turn on the tips of five fingers. The stuffed buns wrapped by Manager Wang Zhengying had 24 folds, which were as sharp as cut by a knife. After being chopped off with a knife from the middle, the opening and bottom of the steamed stuffed buns were the same thick. Master rolling the pieces would produce a piece in three seconds so as to provide enough pieces since so many people were wrapping and his craftsmanship was really superb.

At the end of November 1972, Xiangrui Restaurant was officially opened. Li Qigui left the making of noodle & pastries and followed Master Ai Changrong to stir-fry. Li Qigui had been living in the large dormitory at No. 24 Menkuang Lane since he went to work, which was very close to the unit, so he came early and left late every day. After getting up at 6 o'clock in the morning, he poured the urinal for Master Wang Youwei, leader of the chopping board, and then went to Xiangrui Restaurant for work. He firstly dug out the cookstove pit, chopped wood and gathered the fire for the master. The first cookstove was Master Ai Changrong's, and the second stove was Li Qigui's. After Li

把茶沏上，就等着吃早饭，吃完饭就等"座儿"，一天的炒菜就开始了。

艾师傅是华北楼的老人儿，从公私合营之后就在后厨当负责人，那个时候不叫"厨师长"或"行政总厨"。艾师傅不但刀工技术好，而且炒菜技艺很高超。艾师傅在西河沿大食堂炒菜，马德明是厨师长，他是副厨师长。后来艾师傅就到了华北楼做厨师长，当时已经是前门一带饭庄的名厨师了。等到祥瑞开业，艾师傅就带着李启贵过来了。老师傅炒菜，没有用一个火的，中间是一个正火，两边各有一个次火。正火用来炒菜，左边的次火是爊菜，比如爊着鱼，右边的火上坐着汤，烧肘子或是黄焖鸡。真所谓好厨师要三头六臂，眼观六路耳听八方。一个厨师只能守着一个火干活儿的，当年是找不到工作的。

李启贵看在眼里，记在心上，知道为什么厨师过去叫"勤行"，为什么学徒没有三冬两夏不能出师，为什么厨艺不等于做饭，而是一门可以传承的技艺。启蒙老师艾长荣的言传身教，加上李启贵特有的勤快劲儿、钻研劲儿，教的真教，学的真学，爷儿俩摽着膀子干，很快李启贵的手艺就出落得有模有样。这三个灶怎么使？汤锅怎么用？什么时候加汤？艾师傅手把手地教，包教包会。艾师傅砧板上的活同样厉害。尤其是片，李启贵深得其真传，比如片鲍鱼，这一刀推进去，"啪"地一拉，就是一片，然后用刀尖从左手的大拇指竖往起一撩，一刀一片、一刀一片，行云流水，不阻不滞。

祖国优秀的烹饪文化、烹饪技艺如何承上启下、继往开来，这是摆在老师傅们和李启贵这些青年一代厨师面前共同的任务。四十多年后李启贵还忘不了："艾师傅待我那是一等一的好，他新买了一辆28的飞鸽自行车，连儿子都不让碰，唯独让我休息的时候骑着回亦庄看父母。"

艾长荣最拿手的菜是"浇汁鳜鱼"。后来李启贵在这个菜的基础上研发了"浇汁海上鲜"。主料是鲜活的东星斑一尾，配料是水发海参、鲜活基围虾、鲜活鱿鱼、葱姜、青蒜，调料是食用油、水团粉、酱油、清汤、精盐、蚝油、料酒、糖、醋。将鲜活的东星斑宰杀好，改成均匀的翻刀片，挂上水团粉，用热油炸至外焦里嫩时放入盘中，海参改成抹刀片，鱿鱼切鱼鳃花刀，基围虾去皮，三种配料焯透。把葱切丝，姜切末，青蒜切段，放入碗中，加入水团粉、酱油、清汤、精盐、料酒、糖和蚝油。锅内放入少量油，烧热后烹入碗汁，炒透后下海参、鱿鱼、基围虾，出锅浇在炸好的东星斑上即可。这道菜东星斑外焦里嫩，海参鱿鱼基围虾集海味与海鲜于一镬，鲜香浓郁，色泽亮丽，富有弹牙的质感。

Qigui finished all the work in the morning, he boiled the soup pot and boiled the chicken, duck and pork hock to stew soup. Then he prepared the tea for his masters, and had breakfast before the day's stir-frying work.

Master Ai was a senior chef of the North China Restaurant. He had been the "person in charge" of the kitchen since the public-private partnership. At that time, he was not called "chef" or "executive chef". Master Ai not only had good cutting skills, but also had excellent cooking skills. He was cooking in the big canteen along the West River. Ma Deming was the head chef and he was the deputy chef. Later, Master Ai served as the head chef of the North China Restaurant and he was already a famous chef among restaurants around Qianmen. When Xiangrui Restaurant was opened, Master Ai came with Li Qigui. When stir-frying, Master Ai used more than one cooking fire, there were a main cooking fire in the middle and a secondary cooking fire on each side respectivley. Main cooking fire was used for stir-frying, the secondary cooking fire on the left was used for cooking slowly, such as cooking fish, and the cooking fire on the right was used to make soups, boil elbows or braise chicken. It is true that a good cook should be versatile, and have sharp eyes and keen ears. A cook only working with one cooking fire could not find a job in those days.

Li Qigui kept it in his mind while watching it. He knew why chefs used to be "diligent"; why apprentices could not start their work without several years of apprenticeship; why cooking was not simply making food, but a skill that can be passed on. With the words and deeds of the first instructor Ai Changrong, who presented his authentic skills, coupled with Li Qigui's unique diligence and study, soon Li Qigui's craftsmanship came to an end. How did these three cookstoves work? How to use the soup pot? When would the soup be added? Master Ai taught him hand in hand and thoroughly. Master Ai's skills in the chopping board were equally sophisticated. In particular, Li Qigui learnt his authentic skills. In case of cutting pieces of abalone, pushing forward the the knife and then pulling with a sudden, one piece was made; then using the tip of the knife to lift it vertically from the thumb of the left hand, one piece at a time, flowing smoothly without hindrance or stagnation.

It was a common task for the old chefs and the young cooks such as Li Qigui to connect the past of excellent cooking culture and cooking skills of China with its future. Even after more than 40 years, Li Qigui still cannot forget, "Master Ai treated me particularly well. He bought a new 28-Model Flying Pigeon bicycle and did not even let his son touch it. He only let me ride back to Yizhuang to visit my parents when I was not working."

Ai Changrong's best dish was Mandarin Fish with Sauce. Later, on the basis of this dish, Li Qigui developed Fresh Seafood with Sauce, main ingredient of which was a live plectropomus leopardus, the ingredients included water-fat sea cucumbers, live metapenaeus ensis, live squid, scallion, ginger and garlic sprouts, and the seasonings were edible oil, water powder, soy sauce, clear soup, refined salt, oyster sauce, cooking wine, sugar and vinegar. Kill the live plectropomus leopardus, cut the fish into uniform pieces obliquely, add water powder, fry them in hot oil until it is tender with a crispy crust, put them into a plate, slice sea cucumbers obliquely, cut the squid into fish gill-shaped pattern, peel metapenaeus ensis, and blanch the three ingredients thoroughly. Cut scallion into shreds, ginger into powder and garlic sprouts into sections, put them into a bowl, add water powder, soy sauce, clear soup, refined salt, cooking wine, sugar and oyster sauce. Add a small amount of oil into the pan, heat it, pour it in a bowl of sauce, stir-fry it thoroughly, add sea cucumbers, squid and metapenaeus ensis, and pour it out of the pan on the fried plectropomus leopardus. This dish was tender with a crispy crust. Sea cucumbers, squid and metapenaeus ensis gathered in one wok. It was fresh and fragrant, bright in color and elastic while chewing.

艾师傅另一道拿手菜是"一品肉"。李启贵也学了个全须全尾儿，满宫满调。制作"一品肉"，要精选一块25公分见方的五花肉，厚及一指。刮毛洗净，开水一焯，上色炒糖色，然后再酱一下。酱好后在方子肉皮上切十字花刀，底下肉连着。等肉蒸好了汤㸆浓了装盘的时候，一大块肉仍然方方正正、齐齐整整。拿筷子一夹，跟方块儿肉丁一样，不但肉质熟了而不失其形，而且甘腴肥美、咸甜适口，堪称当时一道名菜。这道菜足够十个人享用。

1972年9月底中日正式建交。1972年底，艾长荣、李启贵承接了一桌日本神秘客人的宴席。李启贵和师傅艾长荣、砧板师傅王有为三人，专门为日本客人做了一桌饭。这桌饭吃的是50块钱的席面，当时海参才一块五一斤，大对虾三块钱一斤，所以怎么能把这50块钱的菜品安排好，李启贵师徒颇费了些心思。

李启贵至今还依稀记得那天的菜谱：4个凉菜都是8个品种半份的双拼菜。随后是四道热菜：头道菜是葱烧海参，二道菜是油焖大虾，三道菜是鲍鱼菜心，四道菜是芙蓉鸡片。汤是烩乌鱼蛋。从菜品色彩安排上颇费心思：葱烧海参是黑的，油焖大虾是红的，鲍鱼菜心是绿的，芙蓉鸡片是白的，烩乌鱼蛋汤是淡黄色的。还上了一道糟熘三白，乳黄色，糟香味浓，口感微甜。随后是香酥鸭子、浇汁鱼，一桌地道的京味鲁菜博得了日本客人的一致好评。李启贵说，当时来了7辆轿车，从前门大街进了廊房二条，饭馆门口都停满了。只是那天他一直在后厨忙活，始终不知道到底参加晚宴的是哪几位日本贵宾。直到今天，李启贵这个谜团还没有解开。

1972年秋末冬初的时候，宣武区针对年轻人搞了一次餐饮业的"技术练兵"。练兵展示的项目很多，有的是切肉片、剁肉丁、拉鸡丝，也有白案的展示项目。其中有一位叫柳振江的小伙子表演抻龙须面。他是致美斋著名面点师董子光的徒弟。柳振江抻的叫"大板儿面"。这是李启贵第一次看抻龙须面，只见柳振江抻完了面，挑着面丝一转圈儿展示，然后把面"啪"地一撒，根根细如发丝，一团面粉纷纷扬扬、洒洒落落，可以说霎时间变成了面团儿，又在人家手中变成了雪花龙须面。李启贵看得目瞪口呆，心里说这太好了，这手活儿我一定要学会！

回到瑞宾楼，李启贵就在面案的案板上天天练这个，每天抻到夜里12点半。那个年代"张铁生交白卷"是潮流，李启贵的苦练被说成是"白专道路业务第一"，受到了瑞宾楼党支部副书记的批判。这位操着山西口音的副书记在

Master Ai's other specialty is "First Class Meat". Li Qigui also learned it authentically. To make "First Class Meat", you should select a 25 cm square pork belly, as thick as one finger. Shave and wash, blanch with boiling water, color by stir-frying sugar, and then add sauce. After that, cut the cruciform pattern on the skin of the square pork belly with the meat attached at the bottom. When the meat was steamed and the soup became thick, it could be loaded on the plate and a large piece of meat was still square and neat. Clip with a pair of chopsticks, just like square cubes of diced meat, which were not only good in its shape, but also appropriate in tastes, and called a famous dish at that time. This dish was enough to serve ten people.

At the end of September 1972, China and Japan formally established diplomatic relations. At the end of 1972, Ai Changrong and Li Qigui received a banquet for mysterious Japanese guests. Li Qigui, his Master Ai Changrong and Chopping Board Master Wang Youwei made a special table for Japanese guests. This table was worth RMB50. At that time, the sea cucumber was only RMB1.5/500g and the prawn was RMB3/500g. Therefore, Li Qigui and his apprentices made efforts to design the dish.

Li Qigui still vaguely remembered the menu of that day: 4 cold dishes were half portions of double dishes in all 8 varieties. Then there were four hot dishes: the first dish Sea Cucumbers with Scallion, the second Braised Prawns with Oil, the third Abalone and Cabbage, and the fourth Lotus & Chicken Slices. The soup was Braised Mullet Egg. The color of the dishes was also given special consideration: the black Sea Cucumbers with Scallion, the red Braised Prawns with Oil, the white Abalone and Cabbage green, the Lotus & Chicken Slices , and the light yellow Braised Mullet Egg. There was also Sauteed Chicken, Fish and Bamboo Shoots with Rice Wine Sauce, which was milky yellow, fragrant and strong flavors, and tasted slightly sweet. Then followed by Crispy Duck and Fish in Sauce. A table of authentic Beijing-style Shandong cuisine won unanimous praise from Japanese guests. Li Qigui said that seven cars entered the Langfang Twotiao from Qianmen Street and occupied the front of the restaurant. It was just that he had been busy in the kitchen that day and did not know which Japanese distinguished guests were attending the dinner. Until today, the mystery of Li Qigui has not been solved.

At the end of autumn and the beginning of winter in 1972, Xuanwu District held a "technical training" for young people in the catering industry. There were many items on display in the training, some were cutting meat slices, chopping diced meat, pulling chicken shreds, and some were making noodles and pastries. Among the displayers, a young man named Liu Zhenjiang, performed to pull Longxu Noodles. He was an apprentice of Dong Ziguang, a famous pastry chef in Zhimeizhai Restaurant. Liu Zhenjiang pulled Wide Noodles. This was the first time for Li Qigui to see the pulling of Longxu Noodles. He saw Liu Zhenjiang drawing out the noodles, carrying the noodles in circles and displaying them. Then he sprinkled the noodles, which were as thin as hair, like a mass of flour drifting profusely and orderly. It could be described that it suddenly turned into dough and snowflake Longxu Noodles in his hands. Li Qigui was dumbfounded and praised in his mind: It's great! And I must learn this skill!

Back in Ruibin Restaurant, Li Qigui practiced this skill every day on the chopping board for making noodles and pastries until 12:30 p.m. every day. At that time, "Zhang Tiesheng handed in blank examination papers" was a trend. Li Qigui's hard training was said to be for the first in professional business and was criticized by a deputy secretary of the Party Branch of Ruibin Restaurant. The deputy secretary with Shanxi accent said at the criticism meeting before the whole store, "Some young people do not sleep in the middle of the night. This is for the first in professional business, which is very inconsistent with the current situation! With the waste of water and electricity, we

全店批判会上说："有的青年人，半夜三更不睡觉，这是白专道路业务第一的苗子，跟当前的形势很不符合！浪费水，浪费电，我们要和他坚决斗争！"后来李启贵犟劲上来了，该练还练，谁也不怕。面抻完了再搋回去，第二天可以接着包饺子、蒸包子，没有造成浪费，就这样练了好一阵子。后来李启贵承认，那个时候抻龙须面，还基本属于蛮练，并不得法，更说不上是特别专业的训练。不过，青年李启贵的汗水并没有白流，人生命运中一个更大的舞台正等着他健步登场。

must resolutely fight against him!" Later, stubborn as Li Qigui was, he continued practices and was not afraid of anyone. After drawing out the noodles, he mixed them together again and could make dumplings and steamed buns without any waste. He practiced for a long time. Later, Li Qigui admitted that at that time, pulling Longxu Noodles was basically a rash practice without knowledge of proper skills, not to mention a special professional training. However, the efforts of young Li Qigui helped him step towards a bigger stage in his life.

1993年，在中国烹饪协会举办的"京、沪、川、粤、有日本参加的两国五方大赛"上，师父王义均正在表演并授艺李启贵大师油爆目鱼花。

In 1993, Wang Yijun was performing and teaching Li Qigui the cooking of Deep Fried Cuttlefish at the "Five-party Competition of Beijing, Shanghai, Sichuan, Guangdong and Japan" held by the China Cuisine Association.

1980年4月16日，北京市饮食公司山东菜系厨师进修班结业纪念合影。
On April 16, 1980, a group photo was taken to commemorate the completion of the advanced training course for chefs of Shandong cuisine of Beijing catering companies.

王义均恩师七十寿诞。
The 70th birthday of Wang Yijun.

1993年，李启贵大师参加由中国烹饪协会举办的"京、沪、川、粤、日本的两国五方大赛"，荣获金奖第一名。
In 1993, Li Qigui won the gold medal in the Five-party Competition of Beijing, Shanghai, Sichuan, Guangdong and Japan held by the China Cuisine Association.

1993年9月2日，李启贵大师荣获中国烹饪协会举办的"京、沪、川、粤、有日本参加的两国五方大赛"金奖第一名。
On September 2, 1993, Li Qigui, one of the top 10 culinary masters, won the first prize in the Five-party Competition of Beijing, Shanghai, Sichuan, Guangdong and Japan held by the China Culinary Association.

第二回

Chapter Two

得偿夙愿 抻龙须三冬两夏
神秘来信 泰丰楼重振雄风

到了1979年初，北京市正好举办一个高级厨师培训班，李启贵到那里深造厨艺。当时北京市饮食公司的一把手是刘峰局长，副局长是王昶，总经理是罗隐恋。总公司的办公地点在河南饭庄的6楼。彼时，李启贵接替已经退休的王正英，在正阳春当上了经理。在培训班上，李启贵和华北楼的副厨师长黄纪明成了互帮互助的好朋友。

有一天，黄继明拉住李启贵，郑重其事地说："启贵，你人真好，我想为你做一件事。改天我领着你到大栅栏施家胡同去串个门儿。"李启贵问他去见谁？黄继明还故意留个扣子，说到了那儿再说。李启贵是个急性子，一个劲儿地追问。黄继明这才交了底："我老丈人叫周子杰，你听说过吗？"李启贵连忙说："我听说过呀！他这个面号称中国一绝，大名鼎鼎，如雷贯耳！"黄继明说："老爷子在抻面上确实有绝活儿，我身子瘦，胳膊没劲儿，学不了。启贵，我看你行！你既有吃苦耐劳的精神，身子骨又结实，能把老爷子的能耐继承下来。你想不想学？"李启贵脑海里又浮现出当年看董子光徒弟柳振江抻龙须面的情景，"学！我当然想学！"

君子言而有信，黄继明真把李启贵带到周子杰家里去了。他跟老爷子说，这是我的同学，我们关系不错，他想跟您学学抻面，您看行不行？周子杰是位老山东，烟台人。张嘴一口胶东腔儿："我的孩子，太辛苦了，你们学不了啊！你没办法，弄不成啊！我岁数也大了，干不了干不了。"老爷子连摆手带摇头，几句话给回了。

这可怎么办？好在黄继明心眼儿活泛，东方不亮西方亮，岳父这儿说不通，找丈母娘说去。黄继明的老岳母也是行里人，在廊坊二条顶头的永生饺子馆包饺子。一听姑爷的央告，老太太当时就冲周子杰发话了："老山东你干什么！孩子要学学，你保守什么？你保守什么！"老伴一发火，周子杰的话锋也软和了，他问李启贵："孩子，你吃得了苦吗？你半路跑不跑？你要跑了，咱们现在干这事，没有意义呀。"李启贵见有门儿，就立马把胸脯子一拔，跟老爷子保证："您老放心，我保证不跑，一定会跟您好好学，一定学会了！"峰回路转，学艺的事儿就这么定下来了。

Realizing the Long-cherished Wish and Pulling the Longxu Noodles for Three Winters and Two Summers
A Mysterious Letter Revitalizing Taifeng Restaurant

At the beginning of 1979, Beijing just held a senior chef training class, where Li Qigui went to further his cooking skills. At that time, Beijing Catering Company had a leadership of Director Liu Feng, Deputy Director Wang Chang, and General Manager Luo Yinlian. The head office was located on the 6th floor of Henan Restaurant. Then, Li Qigui succeeded the place of retired Wang Zhengying as the manager of Zhengyangchun Restaurant. In the training course, Li Qigui and Huang Jiming, Deputy Chef of North China Restaurant, became good friends helping each other.

One day, Huang Jiming took hold of Li Qigui and said solemnly, "Qigui, you are so kind and I want to do something for you. I'll take you to visit somebody in Shijia Lane of Dashilanr some time." Li Qigui asked him who they were going to visit. Huang Jiming deliberately didn't tell him and said he would know upon their arrival. Li Qigui had a quick temper and kept asking the person to be visited. Only then did Huang Jiming come to an end, "My old father-in-law is Zhou Zijie. Have you heard of his name?" Li Qigui hurriedly said, "I have heard of him! His skill in making noodles is unique in China. He is known far and wide." Huang Jiming said, "The old man does have a unique skill in pulling noodles. I am thin and have no strength in my arms so I can't learn it. Qigui, I think you can do it! You have hard-working spirit and strong body so you can inherit the old man's skills. Do you want to learn?" Li Qigui recalled the scene of watching Dong Ziguang's apprentice Liu Zhenjiang pulling Longxu Noodles. "Yes! Of course, I want to learn it!"

The gentleman kept his word. Huang Jiming really took Li Qigui to Zhou Zijie's house. He introduced him to the old man as his classmate and good friends to learn pulling noodles. Zhou Zijie was from Yantai, Shandong. With a Shandong accent, the old man said, "Young man, it's too hard for you to learn! No! I am too old to instruct you." He refused decisively with a few words, waving hands back in the air and shaking head.

How to fix it? Fortunately, Huang Jiming was pretty flexible. He changed his ideas and turned to his mother-in-law, who was also a member of the cooking industry, making dumplings in Yongsheng Dumpling House in Lane II, Langfang. Hearing her son-in-law's plea, the old lady said to Zhou Zijie immediately, "What are you doing, guy? The young man wants to learn some skills. Why not? Why?" When his wife got angry, Zhou Zijie changed his attitude. He asked Li Qigui, "Young man, could you endure the hardship? Would you give up halfway? It's meaningless for us to do this now if you give up halfway." When Li Qigui saw an opportunity, he immediately plucked up his courage and promised, "Please be assured. I promise not to give up. I will work hard and learn from you!" As a result, it was settled.

怎么学呢？李启贵在他住的地方弄了块两米二长、一米二宽的大面板，把周子杰老爷子接来，手把手地教。李启贵下了班就天天在家练抻面，抻好了就把面条送给街坊邻居们吃。那程子，右安门外第三旅馆附近的街坊们可真没少吃李启贵抻的各种面条。

功夫不负有心人，寒来暑往，李启贵抻面的技艺越来越高。周子杰提议在什刹海边上有处房子就在那儿抻面，又在西单的稻香村一进门处租了个柜台，专门卖李启贵和老爷子抻的龙须面。一开张，买主特别踊跃，每天都能卖出120份龙须面。夏天，李启贵索性只穿条三角裤衩，在那儿挥汗如雨地抻面。周老爷子也把自己压箱底儿的制面手艺翻出来，一一传授给李启贵，今天教"一窝丝"，明天教"韭菜扁儿"，后天教"行长条儿"，大后天教"荞麦棱儿"，再后天教"空心面"。李启贵一靠勤学，二靠苦练，三靠灵气儿，足足学了三冬两夏。为什么要学"三冬两夏"呢？周子杰师傅说："冬天的面，你伸手它是又硬又僵，你抻不开；夏天的面跟豆腐渣似的，抓起来往下流，你不练会了这个，没戏。"

身怀抻面绝艺的李启贵，回到正阳春饭庄更加底气十足。此时此刻，他既有艺不压身的喜悦，又有艺无止境的清醒认识。机会总是留给有准备的人，这句话在李启贵的身上应验了。

1983年4月30号中午一点多，时任北京市饮食服务总公司外事处处长兼饮食处副处长的李正权，手里攥着一封信，走进了正阳春李启贵的经理办公室。李启贵赶忙起身让座、沏茶。因为1978年李启贵参加北京市举办的高级厨师培训班时，李正权就是他的班主任，主讲烹饪史理论课。李老师落座后开门见山："启贵，现在北京市正在大力恢复老字号，有这么个事儿，你看你行不行。"说着就把这封信递了过来。李启贵一看，信是从香港寄来的，是写给荣毅仁先生的。这到底是怎么一回事呢？

原来，香港有个老板叫孙必光，过去是北京前门外老泰丰楼的老板，北平光复后，还曾在门框胡同24号开过国民党将校呢军服的加工厂。孙必光来信的意思，是要求恢复老字号泰丰楼饭庄。

荣毅仁之所以把这封信转到李正权手上，一是因为李正权当时是外事处正处长兼饮食处副处长，二是因为李正权既是北京烹饪界的名流，也是一位有着丰富阅历的传奇人物。他是黄埔军校第19期学生，毕业后成了国民党随军记者，新中国成立，他又在餐饮业贡献己之所长。退休后担任中国烹饪协会的副秘书长，几年前以93岁高龄辞世。

How to learn? Li Qigui got a 2.2-meter-long and 1.2-meter-wide board in the place where he lived, and picked up Zhou Zijie to teach him hand in hand. Li Qigui practiced pulling noodles at home every day after work, and gave the pulled noodles to the neighbors to eat. Then, neighbors near the Third Hotel outside Youanmen often ate noodles made by Li Qigui.

Everything comes to him who waits. As time passed, Li Qigui's skill in pulling noodles was becoming better and better. Zhou Zijie proposed to have a house near Shichahai to make noodles, and rented a counter at the entrance of Daoxiang Village in Xidan to sell Longxu Noodles made by Li Qigui and Senior Zhou. As soon as it opened, it was very popular and could sell 120 parts of Longxu Noodles every day. In summer, Li Qigui simply wore only triangular underpants and pulled noodles with sweat. Senior Zhou also turned out his last noodle-making skills and showed them to Li Qigui one by one, including silk-shaped, leek-shaped, long strip, buckwheat ridge-like and hollow noodles. Li Qigui studied hard, practiced diligently and learned for three winters and two summers. Why to have learned for three winters and two summers? Master Zhou Zijie said, "In winter, when you pull noodles, they are hard and stiff, and you can't draw it out. In summer, noodles are like bean curd residue, which is hard to hold. If you don't practice in different seasons, you cannot learn it."

Li Qigui, who was extremely skilled in pulling noodles, was even more confident when he returned to Zhengyangchun Restaurant. At that moment, he had not only the joy of learning more skills, but also clearly knew that skills were endless. Opportunity favors only the prepared mind, which came true in Li Qigui.

At 1:00 p.m. on April 30, 1983, Li Zhengquan, then Director of the Foreign Affairs Department and Deputy Director of the Catering Department of Beijing Catering Service Corporation, walked into Li Qigui's office, the manager's office in Zhengyangchun Restaurant with a letter in his hand. Li Qigui quickly got up, offered him a seat and treated him with tea. Because when Li Qigui attended the senior chef training class held in Beijing in 1978, Li Zhengquan was his head teacher, giving a lecture on cooking history theory. After taking a seat, Mr. Li came straight to the point, "Qigui, Beijing is making great efforts to restore time-honored brands. A matter needs your consideration." As he spoke, he handed the letter over. Li Qigui saw that the letter was sent from Hong Kong and was addressed to Mr. Rong Yiren. What the hell was going on here?

It turned out that there was a boss in Hong Kong named Sun Biguang, who used to be the boss of the former Taifeng Restaurant outside Qianmen in Beijing. After the recovery of Peiping, he also opened a factory processing military uniform of Kuomintang generals at No. 24 Menkuang Lane. Sun Biguang's letter was to request the restoration of the time-honored Taifeng Restaurant.

Rong Yiren transferred the letter to Li Zhengquan because Li Zhengquan was at that time the Director of the Foreign Affairs Department and Deputy Director of the Catering Department, and also because Li Zhengquan was not only a celebrity in Beijing's culinary circle, but also a legendary figure with rich experience. He was among the 19th session of students from Whampoa Military Academy. After graduation, he became a Kuomintang army correspondent. With the founding of New China, he contributed his best to the catering industry. After retirement, he served as the Deputy Secretary-general of China Cuisine Association and died a few years ago at the age of 93.

李正权和李启贵的关系可不一般。李启贵参加"文革"后的北京市首个厨师高级培训班的时候，李正权是他的班主任，主讲中国烹饪史。后来也是通过李正权先生的介绍，李启贵正式拜了丰泽园的鲁菜泰斗王义均大师为师。

恢复泰丰楼的事，李正权和李启贵一拍即合。此后不久，李正权就带着从香港返京的孙必光，上正阳春找李启贵来了。孙必光是个大高个，一看就是典型的老买卖人。双方谈妥后，就开始招兵买马，着手准备恢复老字号泰丰楼。按照李正权的人事安排，李继和任经理，李启贵由经理改任厨师长。经孙必光先生介绍，老泰丰楼掌灶的厨师叫王世珍，全国劳动模范。王世珍就是1952年泰丰楼歇业后，到丰泽园带起王义均厨艺的老师傅。二灶是吴振章，个头儿高，大脚巴丫，外号"吴大脚"。75岁的吴师傅住在果子巷包头章胡同8号。账房先生叫王时光，一个白白净净的老头儿，宽脑门。李启贵见到他时，老爷子已经82岁高龄了。堂头叫孙尚发，时年也七旬开外，家住在如今友谊医院斜对面的黄山酒楼附近。老泰丰楼要重新开张，李启贵把他们都请回来了，老英雄们又有了用武之地。经过一段磨合调整，老师傅们相继功成身退，李启贵又请来了另一位大师级的厨神马德明师傅。

1984年3月8号，泰丰楼正式重张。他们率先恢复的菜品有葱烧海参、红扒熊掌、白扒熊掌、红扒鱼翅、白扒鱼翅、一品锅、油爆双脆、油焖大虾、山东海参、山东菜、松鼠鳜鱼、烧烩爪尖。其口味纯正的京味鲁菜和耐心周到的服务理念，很快为泰丰楼赢得了良好的口碑。更令吃主们惊喜的是，泰丰楼在改革开放之初就恢复了"先用餐后付账"的餐饮业老传统，既体现了对顾客的充分信任，又开了风气之先。这样一来，泰丰楼日益兴隆、宾客如云的热象也就不足为奇了。

提起李启贵和马德明老爷子的缘分，还得从李启贵刚入行说起。当时马师傅在前门的人民餐厅工作，"文革"前这里叫"兴盛馆"。李启贵在祥瑞上班，天天站那儿包包子，能看见马师傅从门口上下班经过。一路过门口，他就过来看看，跟李启贵聊聊。马师傅一点儿也不保守："小李子，你得空儿上我那儿去，我教你做芙蓉鸡片。"马师傅在新中国成立前就是"八大楼"里有名的厨师，有了钱，置了铺面，也成了私方小业主。"文革"时让他改造，曾经到祥瑞饭馆工作过两年多。李启贵天生好学，遇到这么个有手艺的师傅，自然愿意多亲多近。后来就到马师傅家学做菜的能耐，受益良多。

Li Zhengquan and Li Qigui had a close relationship. When Li Qigui attended the first advanced training course for chefs in Beijing after the Cultural Revolution, Li Zhengquan was his head teacher and gave a lecture on the history of Chinese cuisine. Later, with the introduction of Mr. Li Zhengquan, Li Qigui officially took Master Wang Yijun, a leading Shandong cuisine expert in Fengzeyuan Hotel, as his instructor.

Li Zhengquan and Li Qigui totally agreed on the restoration of the Taifeng Restaurant. Shortly thereafter, Li Zhengquan brought Sun Biguang, who returned to Beijing from Hong Kong, to visit Li Qigui in Zhengyangchun Restaurant. Sun Biguang was tall, a typical old businessman at first sight. After the two sides reached an agreement, they began to recruit staff and prepared to restore the time-honored Taifeng Restaurant. According to Li Zhengquan's personnel arrangement, Li Jihe was the manager and Li Qigui was the chef to replace the manager. According to Mr. Sun Biguang, the chef in charge of the kitchen in the old Taifeng Restaurant was Wang Shizhen, a national model worker. Wang Shizhen was the old master who raised Wang Yijun's cooking skills in Fengzeyuan Hotel after Taifeng Restaurant closed down in 1952. The secondary chef was Wu Zhenzhang, who was tall and had big feet and so was nicknamed "Wu Dajiao" (big feet). Master Wu, 75 years old, lived at No. 8 Baotou Zhang Lane, Guozi Alley. Wang Shiguang, a clean old man with a wide forehead, took charge of the account. When Li Qigui saw him, he was 82 years old. The foreman was Sun Shangfa in his seventies living Huangshan Restaurant, which is now diagonally opposite the Friendship Hospital. In order to reopen the time-honored Taifeng Restaurant, Li Qigui had invited them back and the old heroes would play their respective roles. After a period of adjustment, the old masters retired one after another. Li Qigui invited another master chef, Master Ma Deming.

On March 8, 1984, Taifeng Restaurant was officially reopened. The dishes firstly served included Scallion-Roasted Sea Cucumbers, Braised Bear Paw, Grilled Bear Paw, Braised Shark's Fin, Grilled Shark's Fin, First Class Pot, Oil-Fried Double Crisp, Braised Prawns, Shandong Sea Cucumbers, Shandong Cuisine, Squirrel & Mandarin Fish, and Stewed Claw Tips. The authentic Beijing-style Shandong cuisine and patient and considerate service concept soon won a good reputation for Taifeng Restaurant. What surprised the customers most was that at the beginning of the Reform and Opening Up, Taifeng Restaurant restored the old tradition of "eating first and paying the bill later" in the catering industry, which not only reflected the full trust in customers, but also took a lead. As a result, Taifeng Restaurant was becoming more and more prosperous, and received a lot of customers.

The acquaintance of Li Qigui and Ma Deming began from Li Qigui's start of career. At that time Master Ma worked in the People's Restaurant at Qianmen, which was called the Prosperity Restaurant before the Cultural Revolution. Li Qigui worked in Xiangrui Restaurant and made steamed buns there every day. He could see Master Ma commuting before the door. As soon as he passed the door, he came to have a look and talk to Li Qigui. Master Ma was not conservative at all, "Little Li, please come to my house when you have time. I'll teach you how to cook lotus & chicken slices." Before liberation, Master Ma was a famous cook in the "Eight Buildings". When he made money, he set up a store and became a small private owner. During the Cultural Revolution, he was asked to be transformed and then worked in Xiangrui Restaurant for more than two years. Li Qigui was born studious. When he met such a skilled master, he was naturally willing to be more close to him. Later, he went to Master Ma's house to learn how to cook and benefited a lot.

十多年后，泰丰楼的重张又把这投缘的师徒俩粘到了一起。据李启贵讲，马师傅来泰丰楼掌灶，还和泰丰楼有个特殊而幽默的约定。只要马师傅来店里，想喝酒，启贵就负责打酒，别人打不行。马师傅白喝不要钱，这叫"吃柜上喝柜上"，特殊的优待。跟着马师傅，启贵也如鱼得水，厨艺日益精进。他跟马师傅学的名菜包括：通天鱼翅、清汤官燕、芙蓉官燕、兰花熊掌、糟熘三白、糟熘鸭肝、烩乌鱼蛋、三吃丸子、芫爆肚丝……马师傅最拿手的菜还有"游龙戏凤"、四吃活鱼、红烧海参、干烧冬笋、油爆双脆、芙蓉鸡片。"游龙戏凤"的菜名显然是借用了同名京剧的噱头，主料是大对虾和鸡里脊。配料是鸡蛋、葱姜、面粉、莴笋等，调料要用到精盐、清汤、姜汁、水团粉、胡椒粉、葱姜油。制作方法并不复杂，关键看火候的掌握。先将鸡里脊去筋，花刀切条状，加入面粉、鸡蛋，和水团粉抓糊炸好放盘中间。再将大虾处理好，从背部开刀翻过来，加料入味儿，上屉蒸好，码在鸡里脊条的两边，中间用莴笋刻成门墩丁，汤锅上火放入清汤，加入精盐、姜汁、胡椒粉勾芡，淋入葱姜油，浇在大虾上即可。这道菜吃起来，大虾海鲜味足、形状美观、色泽红润，鸡里脊条软嫩适口、老少咸宜。

"芫爆双脆"与"游龙戏凤"有着异曲同工之妙。"芫爆双脆"的主料是鲟鳇鱼的鱼信（鱼筋）和麦穗蚕头。配料是芫荽（香菜）、青蒜、葱丝、蒜、姜。调料包括精盐、胡椒粉、料酒、醋、清汤、糖、香油等。先把鲟鱼筋用剪刀划开去心，洗净，切成条状寸段备用。然后把麦穗蚕头的根去掉，葱切丝，姜切末，蒜一半儿切片一半儿切蒜蓉。香菜切寸段备用。然后汤勺上火放水，将鲟鱼筋和麦穗蚕头焯透，香菜放入碗中，加入葱丝、姜末、蒜蓉、蒜片、胡椒粉、料酒、醋、清汤、糖备用。最后炒锅上火烧热油，放入两种主料和配料翻锅拌匀，点香油，装盘即可。这是京味鲁菜的"四大爆"之一，清淡爽脆，鲜香适口，白绿相间。

马师傅教启贵"长燕窝"，长燕窝就是水发燕窝，夏天用凉水，春秋天、冬天用点儿温水，稍微热一点。燕窝先泡，泡开了之后，拿一个干净水碗，往出择毛儿，择完了在干净的水碗里涮。走燕窝的时候，还得拿碱长。这个需要五六分钟，碱就合适了。然后开始撤碱。撤碱的时候用的开水不能低于三遍。撤一遍，燕窝一长；撤一遍，燕窝一长。最后用干净的小毛巾把水挤出去，燕窝就发好了。

More than ten years later, the re-opening of Taifeng Restaurant linked the instructor and apprentice together, who were congenial. According to Li Qigui, Master Ma came to Taifeng Restaurant with a special and humorous agreement with Taifeng Restaurant. As long as Master Ma came to the restaurant and wanted to drink wine, Qigui would be responsible for preparing liquor, but others couldn't. Master Ma didn't pay money to drink, which was a special preferential treatment. Following Master Ma, Qigui felt just like a fish in water, and his cooking skills were getting better and better. The famous dishes he learned from Master Ma included Super Shark's Fin, Fine White Cubilose in Clear Soup, Fine White Cubilose with Lotus, Orchid & Bear's Paw, Sauteed Chicken, Fish and Bamboo Shoots with Rice Wine Sauce, Sauteed Duck Liver with Rice Wine Sauce, Stewed Black Fish Eggs, Delicious Meatballs, Sauteed Sliced Pork Tripe with Coriander ... Master Ma's best dishes included Dragon and Phoenix, Delicious Live Fish, Braised Sea Cucumbers, Dry-fried Winter Bamboo Shoots, Oil-fried Double Crisp and Lotus & Chicken Slices. The dish name of "Dragon and Phoenix" obviously borrowed the gimmick of Peking Opera with the same name. The main ingredients were prawn and chicken tenderloin. Ingredients included eggs, scallion and ginger, flour, asparagus lettuce, etc. Seasonings included refined salt, clear soup, ginger juice, water powder, pepper and fried scallion-ginger oil. The making method was not complicated, and the key depended on the control of duration and degree of heating. First, remove the tendons from the chicken tenderloin, cut it into strips crosswise, add flour, eggs, and water powder to paste, then fry and put it in the middle of the plate. Then handle the prawns, cut from the back and turn them over, add seasonings, steam them in a drawer, and place them on both sides of the chicken tenderloin strips. The asparagus lettuce was carved into cubes in the middle. Heat the soup pan, pour clear soup, add refined salt, ginger juice and pepper to thicken, and pour fried scallion-ginger oil into the prawns. This dish tastes appropriate at prawns and seafood, with beautiful shape, ruddy color, soft and tender chicken tenderloin strips and palatable taste for customers of all ages.

"Stir-fried Double Crisp with Coriander" and "Dragon and Phoenix" had the same effect. The main ingredients of Stir-fried Double Crisp with Coriander were the fish tendon and wheatear-shaped jellyfish head. Ingredients were coriander, leeks, shredded scallion, garlic and ginger. Seasonings included refined salt, pepper, cooking wine, vinegar, clear soup, sugar, sesame oil, etc. First, cut the sturgeon tendon with scissors to remove the heart, wash it, and cut it into strip-shaped one-inch sections for later use. Then remove the root of the jellyfish head, shred the scallion, chop the ginger into powder, slice the garlic in half and mince the garlic in half. Cut coriander into one-inch sections for later use. Then heat soup spoon, add water, blanch the sturgeon tendon and jellyfish head thoroughly, put the coriander into a bowl, and add shredded scallion, minced ginger, minced garlic, garlic slices, pepper, cooking wine, vinegar, clear soup and sugar for later use. Finally, heat the oil on the frying pan, add the two main ingredients and ingredients, turn upside down, mix well, sprinkle sesame oil and put it on a plate. This was one of the "four stir-frying dishes" of Beijing-style Shandong cuisine. It tasted light, crisp, and delicious with white and green color.

Master Ma taught Qigui to make "Long Bird's Nest". Long Bird's Nest was to water-fat bird's nest. Use cold water in summer and warm water in spring, autumn, and winter, which should be slightly heated. Soak bird's nest first, then take a clean water bowl, pick up plume, and rinse in the clean water bowl. When making bird's nest, alkali would be used to make it larger, which would take about five or six minutes, and then begin to sprinkle alkali for no less than three times with the boiling water used. Each time, the bird's nest would become longer after the boiling water was changed. Finally, use a clean towel to squeeze out the water and the bird's nest would be ready.

马师傅教启贵发鱼翅，过去都讲究通天鱼翅、披刀翅、金钩翅、荷包翅。李启贵回忆，马师傅教他发海参最为细致。干海参首先要洗，洗完了泡，泡完了煮，煮好要用保鲜纸封上闷起来。闷完了有硬的要挑要筛选。筛选后，硬的还得要煮，必须要用纯净水，海参里有盐，换一遍水，撒一遍盐，海参一长。最后要拿天然冰"追"，一"追"这海参就全起来了。然后根据海参的种类特点进行烹制。日本海参刺儿尖，但是肉脆。大连海参刺儿稀，刺儿不那么尖，肉也厚实。海参崴的海参呢，不但肉厚、个儿大，刺儿也好。烹制的方法都不一样。

李启贵跟马师傅学的拿手菜叫"红扒通天鱼翅"。这道菜的主料是披刀翅一块。配料是鸡鸭肉、肘子、干贝、火腿。调料是清汤、料酒、精盐、水团粉、葱姜油、鸡油、葱姜。先将披刀翅发透，放入砂锅中，加入葱姜、料酒蒸透，滗去原汤，再把鸡鸭、瘦肉、肘子斩成块儿，焯水，鱼翅用细纱布包好，上面再放一层鸡鸭、肘子、瘦肉。汤锅上火放油，烹入葱姜，对入清汤、料酒、精盐烧开，撇去浮沫，浇在鱼翅上，用旺火烧开，再改为小火煨。煨至鱼翅软烂但不失其形，放入汤锅内，加入清汤和干贝汤、料酒、酱油提色，烧开，用水团粉勾芡，放入葱姜油、鸡油，大翻锅，装盘撒火腿米即可。这道菜的特点是鱼翅软烂而不失其形，翅香滑润而汁浓味厚。

做这些看家菜的每一个细节，马师傅都仔仔细细、原原本本地教给了李启贵，每天早上上班前，还让启贵到家里上"一对一的培训班儿"。爷儿俩笃厚的情谊，让同行里不少人心生羡慕。直到1985年底，李启贵代表中国奔赴欧洲，作为中国代表团的第一主力参加世界奥林匹克烹饪大赛，马师傅才功成身退。但是二人的忘年之谊从未间断。

在泰丰楼，李启贵的厨艺精益求精，始终没有把学习和基本功扔在一边。1984年李启贵在泰丰楼工作的时候，宣武区烹饪协会办了一个烹饪班，有四十多名学员，李启贵负责讲热菜的课。刀功课由另外一位老师讲授。没想到，刀功课上的一个学习内容把讲课的老师难住了。什么内容呢？教学现场抬来一扇猪肉，没去皮、没去骨，要求指认猪身上的所有部位名称，并写成小纸条，贴在相应的部位上，然后把各部位依次选料取下，再按原来位置摆回去。那位授课老师端详了一阵子，又抬眼看了看学员们，发现学员们也正盯着他。老师脑门子冒汗了，找到班主任悄悄说，这个我弄不了，赶紧找能人吧。班主任立马拨通了泰丰楼经理的电话："让启贵来救场吧！"

Master Ma taught Qi Gui to water-fat shark's fin. In the past, it's popular to use super shark's fin, knife-like shark's fin, gold hook-like shark's fin and purse-like shark's fin. Li Qigui recalled that Master Ma taught him the most meticulous way to water-fat sea cucumbers. Dried sea cucumbers should be washed first, soaked after washing, boiled after soaking, and sealed with fresh-keeping film before simmering. After simmering, pick up and screen the hard ones. After screening, boil the hard ones. Pure water must be used. Since sea cucumbers have salt, change the water to remove the salt repeatedly, and the sea cucumbers would grow longer. Finally, use the natural ice to cold the sea cucumbers so as to make them stand up. Then cook according to the species and characteristics of sea cucumbers. Japanese sea cucumbers have sharp thorns but crisp meat. Dalian sea cucumbers have thin but not sharp thorns and thick meat. The sea cucumbers in Vladivostok are not only thick in meat, but also big and thorny. The Steps are different.

The specialty Li Qigui learned from Master Ma was called "Braised Super Shark's Fin". The main ingredient of this dish was a piece of knife-like shark's fin. Ingredients were chicken and duck meat, pork elbow, dried scallops and ham. Seasonings were clear soup, cooking wine, refined salt, water powder, fried scallion-ginger oil, chicken oil, scallion and ginger. First, water-fat knife-like shark's fin thoroughly, put it into a casserole, add scallion, ginger and cooking wine, steam them thoroughly, decant the original soup, then cut the chickens, ducks, lean meat and pork elbows into pieces, blanch them, wrap the shark's fin with spun yarn cloth, and put a layer of chickens, ducks, pork elbows and lean meat on top. Add oil to the soup pan and heat it, cook scallion and ginger, add clear soup, cooking wine and refined salt to boil, skim off floating foam, pour on shark's fin, boil with strong fire, and then change to stew with weak fire. Until having soft and easy-to-eat shark's fin without losing its shape, put it into a soup pot, add clear soup, dried scallops soup, cooking wine and soy sauce to improve color, boil, thicken with water powder, add scallion, ginger oil and chicken oil, turn upside down the pot, and sprinkle diced ham on a plate. This dish is characterized by soft and easy-eat shark's fin without losing their shape, tasting smooth with thick sauce.

Master Ma taught Li Qigui every detail of making these specialties carefully and exactly. Before going to work every morning, he also asked Qigui to attend "one-to-one training class" at home. The sincere friendship between them had made many people in the same trade envious. It was not until the end of 1985 when Li Qigui went to Europe on behalf of China to participate in the IKA as the first main force of the Chinese delegation that Master Ma retired. However, their friendship had never stopped.

In Taifeng Restaurant, Li Qigui kept improving his cooking skills and never put his study and basic skills aside. When Li Qigui was working in Taifeng Restaurant in 1984, Xuanwu District Cooking Association organized a cooking class with more than 40 students. Li Qigui was responsible for teaching hot dishes. The cutting skills lesson was taught by another teacher. Unexpectedly, one of the learning contents in cutting skills puzzled the teacher. What was it? A piece of pork was carried to the teaching site without peeling or removing bones. The requirement was to identify the names of all parts of the pig, write a small piece of paper, stick it on the corresponding part, then remove the materials from each part in turn, and put it back according to the original position. The teacher studied for a while, then looked up at the students, finding they were staring at him. The teacher's forehead was sweating. He found and whispered to the head teacher that he couldn't do it, and then asked to find someone who could do it quickly. The head teacher immediately dialed the telephone of the manager of Taifeng Restaurant, "Let Qigui come here and give a hand!"

李启贵走完宴席，开车就赶过去了。到了那儿，提起整扇猪肉来，啪啪啪，连讲带选粘纸条，把各个部位依次分开，猪皮挺下，然后把皮铺在案上，又把这个十几个部位原样对上。当时学员班的班长刘建洲站在旁边不错眼珠地看，只见李启贵右手提刀，左手一把抓住前臀尖，刀贴在后臀尖，啪！这么一掉个儿一转，那个利索劲儿就甭提了。刘建洲心想：这师傅厉害！等李启贵把肉选完了，他凑上前说："师傅您是有真功夫的人啊，我得拜您为师！"后来还真如愿拜了李启贵大师。从1984年开始到现在，师徒俩一直关系很密切。

那么，李启贵的这手剔肉好刀工是什么时候学的呢？那还得从他在瑞宾楼上二灶炒菜的时候说起。当时他认识一个在西河沿菜市场里卖菜的何师傅。他家住在廊坊二条，跟瑞宾楼离着不远。李启贵经常来买带骨头的冻肉，看见来一车冻肉，小何他们几个人忙不过来，年轻的李启贵就利用工休和午休时间来西河沿菜市场帮忙，也是跟着学剔肉选料。剔肉分两种，一种是"大掀盖儿"，就是把排骨整个取下来，用刀一圈，剔肉一开始，上去俩手指，伸手一抓，先把腰子抓下来。啪啪划一个"十"字，把板油摘下来了。然后把飘着的碎肉一拉，提搂下来了。拿刀一圈，从血脖（猪颈肉）开始，从通脊往后到排骨的下边，一圈儿圈完。两手指头拎起来，拿刀贴着骨头走，这是"大掀盖儿"。第二种是"挑单根儿"，在骨头上"唰"地一刀，往外一翻，"啪啪啪"这几个都挑完了。拿刀往上一顶，这单根骨头就剔出来了。菜市场的小何是专业剔肉的，天天一剔能剔好几百片猪。李启贵的剔肉手艺就是跟他学出来的。后来李启贵找来一把废弃的旧钢锉，自己用砂轮磨了一把仰脸刀。钢口之好，胜强百倍，用这个剔骨选料，得心应手。

在瑞宾楼，李启贵的案子边上插着三把快刀，都是他自己找旧刀、锵了刃儿的破刀磨出来的。他天天把肉选好后，剩下的筋头巴脑留着，每天晚上十点左右下班开始练刀工，切到夜里12点半，每天切三大桶，一桶四十多斤，第二天绞馅儿包包子、做饺子都省事儿了。最后练到一小时可以切36斤，一时无双。

有道是"人怕出名猪怕壮"，李启贵的刀工名声在外，引来了要"以武会友"的挑战者。民族饭店一位姓朱的年轻厨师，又高又壮，背着刀拿着肉，上瑞宾楼找李启贵比试来啦。

李启贵一来年轻气盛，二来心里有底，说："好啊。你先来我先来？你先来吧！"他说："那你先来吧。我看看。"李启贵说："那好，你是客人，

Li Qigui drove there after having prepared the banquet. When he got there, he lifted the piece of pork, popped it, labelled while introducing each part. He separated all parts and removed pigskin. Then he spreaded the pigskin on the board and aligned the more than a dozen parts as they were. Then, Liu Jianzhou, the monitor of the student class, stood beside him and looked at him with staring eyes. He saw Li Qigui holding the knife in his right hand and grabbing the front hip tip in his left hand. Closed the knife to the back hip tip and finished after a turn, with the fabulous agility. Liu Jianzhou thought to himself, this master was great! When Li Qigui had finished selecting the pork, he leaned forward and said, "Master, you have real skills. I am eager to be your apprentice!" Later, he really took Master Li Qigui as his master. Since 1984, they have been keeping closely relationship.

So, when did Li Qigui learn this excellent cutting skills? He started when he was cooking at the second cookstove in Ruibin Restaurant. At that time, he knew Master He who was selling vegetables in the vegetable market along the West River. He lived in Lane II, Langfang, not far from Ruibin Restaurant. Li Qigui often came to buy frozen meat with bones. When there was a truck of frozen meat, Little He and his assistants couldn't make it. Young Li Qigui took advantage of his work and lunch break and gave them a help at the vegetable market along the West River. He also learned to pick meat and choose materials. There were two kinds of scraping pork off bone, one by taking off the whole ribs and using a knife in a circle manner. At the beginning of scraping, take off the kidney first with two fingers. Cut crosswise and took off the suet. Then he pulled the floating minced meat and carried it down. Take it in a circle manner with a knife, starting from the blood neck (pig neck meat), through the ridge back to the bottom of the ribs, and finish it in a circle. Pick it up with two fingers and close a knife against the bone, which was the first kind of removing all. The second was to" scrape single bones", swish a knife on the bone, turn it out, and finish in an instance. Take a knife and push the bone up, and the single bone would be removed. Little He in the vegetable market specialized in scraping meat and could scrape hundreds of pigs every day. Li Qigui learned the scraping skills from him. Later, Li Qigui found an old discarded steel file and grinded a face-lifting knife with a grinding wheel. The steel mouth was a hundred times better than the other. It was handy to use this boning material.

In Ruibin Restaurant, on the side of Li Qigui board was three sharp knives, all of which were sharpened by himself from old knives and blade-broken knives he found. After he chose the meat every day, he kept the rest of the remains. He started to practice scraping at about 10 o'clock every night and scraped until 12:30 at night. He scraped three large barrels a day, one of which weighed more than 20 kg, so that it's easy to to mince the stuffing for making steamed buns and dumplings in the next day. In the end, he could scrape 18kg in one hour, which was unparalleled at the moment.

There is a saying that "fame brings a person trouble just as fatness does for pigs". Li Qigui was well-known for his excellent cutting skills, which had attracted challengers who wanted to "meet friends with cutting skills". A young chef surnamed Zhu from the Minzu Hotel, tall and strong, carrying a knife and meat, went to Ruibin Restaurant and wanted to challenge Li Qigui.

On the one hand, Li Qigui was young and vigorous and on the other hand, he was confident and said, "Ok. Who will be first? You come first!" The young chef said, "You come first. Let me see." Li Qigui said, "Well, you are a guest. Look at me first." In the meat-cut-opening room, both of them

你先看我的。"到了开生间,俩人把肉都摆出来。李启贵拿出刀来案前一站,一绷劲儿,唰唰唰,刀都快飞起来了,速度之快到了目不暇接的地步。切片讲究"一刀筒",这一个"肉跑儿",不停刀,一口气就切完了。

小朱师傅一看这个,说:"甭比了,我甘拜下风。"后来才知道,李启贵剔骨按部位选肉挺皮,10分钟一扇肉,这个速度,至今谁也没有达到。

有这样的硬功底儿,李启贵在考一级厨师的时候连拿五个第一,获得考级第一名。李启贵大师回忆说:"考一级厨师,要考五项,第一是开菜单,第二是理论知识,第三是刀工。刀工又分为三小项,一个是拉鸡丝,一个是脱骨鸭,还有一个是葡萄鱼。这条鱼切出来,软边和硬边分出两串'葡萄'来。很多拉不好的,就是刀口拉反了,一嘟噜'葡萄珠儿'朝上长着,另一嘟噜'葡萄珠儿'朝下长着。脱骨鸭子,就是说从鸭脖子后边开一寸五的口子,把鸭子的整个骨头脱出来,鸭子皮不能破。最后通过称分量鉴定,不能鸭子脱骨以后,就剩下鸭子皮没东西啦。第四项是冷菜的图案拼板,第五项是热菜的烹炒,现场炒四个菜。我这五项当时都是第一。"

作为一名在北京烹饪界崭露头角的优秀青年厨师,如何在国门刚刚开启的时代,把中国的烹饪文化从容自信地向世界展示出来?是突然降临到李启贵头上的一道难题。让他始料未及的是,在接下来的一年时间里,从准备大赛到参赛时紧急情况应急处置,一个个棘手问题纷至沓来,考验着包括团长和翻译在内的五人代表团,更考验着挂头牌的主力夺金手李启贵。这个前无古人的任务,李启贵能完成吗?

1985 年,北京市劳动局技师考核汇报。
In 1985, Beijing Labor Bureau held an assessment debriefing for technicians.

spread the meat. Li Qigui took out the knife to the front stop of the board. In one vigorous effort, he was flying the knife at super-speed. Slicing should be finished with one cutting, so he finished cutting in one breath.

Little Zhu looked at this and said, "I give up. I'm willing to give in." Only later did he know that Li Qigui scraped according to the parts of pig, and it took 10 minutes for him to scrape one piece of meat. So far, no one has reached this speed.

With such skills, Li Qigui won the first place in the first-class chef examination with five firsts when he took the first-class chef examination. Master Li Qigui recalled, "To take the first-class chef examination, one has to take an examination of five items. The first is to open a menu, the second is about theoretical knowledge, and the third is cutting skills. Cutting skills are divided into three small items, making chicken shreds, boning duck, and making grape-shape fish. The fish is cut into two bunches of 'grapes' on the soft and hard sides. Many failures are because the knife edge is pulled in a wrong direction, with a bunch of grapes growing upward and the other growing downward. Boning duck is to make a 1.5 inch hole from the back of the duck's neck and take off the whole bone of the duck, and the duck's skin cannot be broken. Finally, weigh to identify whether there is enough meat after being boned except the duck skin. The fourth item is the pattern of cold dishes, and the fifth item is the cooking of hot dishes, with 4 dishes fried on the spot. I was the first in all five items at that time."

As a young outstanding chef who has made his mark in Beijing's culinary circle, how could he show China's culinary culture to the world calmly and confidently in the era when the country just started the opening up? It is a difficult problem that suddenly befell Li Qigui. To his surprise, in the following year, from the preparation of the competition to the emergency response during the competition, thorny problems came one by one, and tested the five-member delegation, including the head of the delegation and the interpreter, and also tested Li Qigui, the main gold medalist who won the first prize. Could Li Qigui complete this unprecedented task?

1986年12月，李启贵大师和中国代表团团长蔡敬之、翻译文彩霞向世界烹协秘书长介绍中国菜肴。

In December 1986, Li Qigui ,Cai Jingzhi and Wen Caixia, head of the Chinese delegation and interpreter, introduced Chinese cuisine to the Secretary General of the World Restaurant Cuisine Association.

1987年，李启贵大师和师父马德明一起到大兴县讲授烹饪课程。
In 1987, Li Qigui and his mentor Ma Deming went to Daxing County to teach cooking courses together.

1994年，泰丰楼第二次装修完开业合影。
In 1994, a group photo was taken after the second renovation of Taifenglou.

2000年11月2日，李启贵被评为十佳烹饪大师。
On November 2, 2000, Li Qigui was awarded as one of the top 10 culinary masters.

2012年2月13日，李启贵大师在美国纽约表演中华八珍宝鼎和雪花龙须面。
On February 13, 2012, Li Qigui performed the cooking of the Chinese Eight Delicacies Banquet and Snowflake Longxu Noodle in New York, USA.

第三回

Chapter Three

雪中送炭 竞技场因谁得救
勇夺首金 开幕式为他推迟

1984年秋天，一封邀请函把李启贵推到了世界烹饪舞台的中央，这是上苍的垂青还是自身的实力？抑或兼而有之？总之，李启贵的名字第一次与中国紧密联系在一起。这封神秘的邀请函是从中国对外友协转到北京市饮食服务总公司的，当时中国烹饪协会还没有成立。

1984年11月18日，在李启贵工作的泰丰楼二楼，北京市饮食服务总公司就此召开了第一次专题会。36年过去了，89岁的北京市饮食服务总公司高仁处长对这个会仍记忆犹新。据高仁先生回忆，这封即将于1985年12月在卢森堡举办的第五届奥林匹克世界烹饪大赛的邀请函，是驻外使馆转给对外友协的。此事之所以引起各方面的高度重视，是因为新中国成立后，还没有任何一个烹饪代表团代表中国参加如此规格的世界顶级赛事。而这一全球最大的赛事早已被各国厨师所熟知，已成为全世界厨师、面点师最期望参与竞技的国际赛事。

高仁先生说："当时咱们接到任务，是对外友协找到了北京市饮食服务总公司，具体有我参与落实。这个赛事由几十个国家和地区的77个代表团参赛，规模在当时是空前的。我们当时根据对外友协的意见，邀请上海、沈阳、广州、天津的同仁们共同参与，他们却纷纷婉言谢绝，表示压力太大不愿参加，怕拿不回金牌，愧对国人。"

于是奇怪的一幕出现了：所谓的"中国代表团"，其实只有"北京代表队"参加。而驻外使馆和对外友协讲，这是新中国成立以来，中国代表团第一次参加世界烹饪大赛，是代表国家参赛，不能以一个城市的名义参加，所以将北京代表队破格提升为中国代表团。团长是北京市饮食服务总公司总经理蔡敬之，翻译由北京马克西姆餐厅的英文、法文翻译文彩霞担任。参赛队员由李启贵、陈守斌、张志广三位厨师组成。别看只有三人出马，却个个身手不凡。时年50岁的陈守斌在全聚德已经大名鼎鼎，技术精湛，水平高超，成为北京烤鸭技艺的代表性人物。张志广亦是鲁菜名家，尤擅冷拼和食雕，他的冷拼作品意趣脱俗、栩栩如生，堪称北京烹饪界的冷拼第一人。李启贵当时不仅年富力强，而且技艺全面，"炸、爆、烧、炒、熘、煮、氽、涮、蒸、炖、煨、焖、

Timely Assistance Rescuing the Competition
Winning the First Gold Medal, He Made the Opening Ceremony of the IKA Delayed

In the autumn of 1984, an invitation letter sent Li Qigui to the center of the world cooking stage. Was this the favor of God or its own strength? Or both? In short, Li Qigui's name was closely linked with China for the first time. The mysterious invitation was transferred from the Chinese People's Association for Friendship with Foreign Countries to the Beijing Catering Service Corporation, and the China Cooking Association had not yet been established at that time.

On November 18, 1984, on the second floor of Taifeng Restaurant where Li Qigui worked for, Beijing Catering Service Corporation held its first special meeting on this issue. Thirty-six years later, Gao Ren, 89 years old, Director of Beijing Catering Service Corporation, still remembered the meeting vividly. According to Mr. Gao Ren's memory, this letter was the invitation for the 5th IKA to be held in Luxembourg in December 1985 transferred to the Chinese People's Association for Friendship with Foreign Countries by embassies abroad. The reason why this issue had attracted great attention from all sides was that since the founding of New China, no cooking delegation had represented China in the world's top competitions of such scale. The world's largest competition had long been well known by chefs from all over the world and had become the international competition that chefs and pastry chefs from all over the world expect to participate in most.

Mr. Gao Ren said, "The task we received at that time is given by the Chinese People's Association for Friendship with Foreign Countries via the Beijing Catering Service Corporation. I take charge of its implementation. The competition is attended by 77 delegations from dozens of countries and regions, with unprecedented scale. According to the opinions of the Chinese People's Association for Friendship with Foreign Countries, we invite colleagues from Shanghai, Shenyang, Guangzhou and Tianjin to participate in, but they decline politely one after another, saying that they are under too much pressure to participate. They are afraid of not winning the gold medal back and would feel ashamed of the Chinese."

Then there was a strange scene: The so-called "Chinese delegation" was actually only represented by the Beijing delegation. This was the first time that the Chinese delegation participated in the IKA since the founding of the People's Republic of China. It represented the country and couldn't participate in the competition under the name of a city, so the Beijing delegation would be elevated unconventionally to the Chinese delegation. The delegation was led by Cai Jingzhi, General Manager of Beijing Catering Service Corporation, and the English and French interpreter was Min Caixia from Maxim's in Beijing. The contestants consisted of Li Qigui, Chen Shoubin and Zhang Zhiguang, and all had extraordinary skills. At the age of 50, Chen Shoubin was already well-known in Quanjude Restaurant for his superb technique and excellent level, and became the representative figure of Beijing roast duck technique. Zhang Zhiguang was also a famous chef in Shandong cuisine, especially good at assorted cold dishes and food carving, and his assorted cold dishes were free from vulgarity

烩、扒、焗、煸、煎、塌、卤、酱、拌、炝、腌、冻、糟、醉、烤、熏"的烹饪二十八法，可谓般般拿手，样样精通。有了这"烹坛三剑客"，中国代表团自然底气十足。

强大的参赛阵容有了，拿什么菜品参赛呢？

京城烹饪界的领军人物们在泰丰楼共商大计。其中包括北京市第一任烹饪协会副会长蔡启厚、全聚德的田文宽、陈守斌大师、丰泽园的王义均大师、萃华楼的丛培德、孙仲才大师、仿膳的董世国大师、泰丰楼的马德明大师、北京饭店的黄子云大师、张志国大师和康辉大师，再加上泰丰楼的"头牌"李启贵大师，可谓集一时之盛。大师们的任务就一个：研究参赛的菜谱。

根据奥林匹克第五届世界烹饪大赛组委会的要求，每个参赛代表团要做一个16平方米的展台，在展台当中要做10个大菜，10个大菜，其中包括两个凉菜、八个热菜和两个现场销售的菜品。现场销售的菜品每道菜要做130份，两道菜共260份。这也是那届大赛评分的一个重要依据。就是说大赛开始之际，各代表团要同时售卖这两个菜，口味最佳、最先售罄者获得高分。这些都需要事先设计好方案，才能在大赛中做到人无我有，人有我优。这样的"神仙会"前后开了三次，菜品方案才初步确定下来。

初步方案敲定，开始演练。地点就选在了和平门全聚德烤鸭店的二楼，在这个宽敞的大厅里搭起了16平方米的大展台。从1985年春天开始，李启贵就和陈守斌、张志广在这里实际操练起来，力求做到胸有成竹，游刃有余。几个月的时间，他们试制了很多菜，有些并不很成功，也有的不具备现场表演的观赏性，不讨巧，有些在操作上耗时太长，不适宜比赛的节奏。于是边试边调整，不断淘汰菜品，不断更新菜品。七八个月的工夫，他们就像实验室里的科学家，不厌其烦地反复实验着，为新中国烹饪技艺在国际舞台的第一次亮相，做着不懈的努力。

他们最终确定的方案是，在16平方米的大展台中间，制作大型食品雕刻作品"天女散花"，其周围是冷拼"喜鹊登梅"和"龙凤呈祥"。八个热菜分别是：宝参冬瓜盅、北京烤鸭、八宝涮锅、百花珍珠鱼、荔枝鸭卷、猴头菇燴大虾、一品芙蓉、父子龙虾。现场制作销售的菜肴有两道：糖醋鱼条、香酥鸭腿。

因为要去欧洲比赛，当地的一些风俗礼仪一定要了解，饮食习惯也要有所掌握。比方说，到欧洲的日子里，要习惯喝冰水、用刀叉、吃牛排。这对

and lifelike. He could be called the first person of assorted cold dishes in Beijing culinary circle. Li Qigui was not only in the prime of his life, but also had comprehensive skills, i.e., 28 cooking skills such as quick-frying, stir-frying, roasting, frying, boiling, quick-boiling, blanching, steaming, simmering, stewing, and braising", all of which were perfectly practiced. The three swordsmen in cooking, the Chinese delegation was naturally confident.

With a strong team, what dishes would they take to participate in the competition?

Leading figures in Beijing's culinary circle discussed plans in Taifeng Restaurant. Among them were Cai Qihou, the first vice-chairman of Beijing Cooking Association, Tian Wenkuan and Chen Shoubin of Quanjude Restaurant, Wang Yijun of Fengzeyuan Hotel, Cong Peide and Sun Zhongcai of Cuihua Restaurant, Dong Shiguo of Fangshan Restaurant, Ma Deming of Taifeng Restaurant, Huang Ziyun, Zhang Zhiguo and Kang Hui of Beijing Hotel, together with Master Li Qigui, the leading chef of Taifeng Restaurant. The team could be described as a gathering of the then bests. The masters only has one task, that's, to study the recipes for the competition.

According to the requirements of the Organizing Committee of the 5th IKA, each participating delegation would make a 16-square-meter booth, in which 10 major dishes should be made, including 2 cold dishes, 8 hot dishes and 2 dishes sold on the spot. The dishes sold on the spot need to be 130 pieces for each dish and 260 pieces for the two dishes in total. This was also an important basis for scoring the competition. In other words, at the beginning of the competition, all delegations would sell the two dishes at the same time. The one with the best taste and the first to sell out would get a higher mark. All of these needed to be designed in advance so as to achieve the goal of being advantageous and excellent in the competition. Such meeting was held three times successively before the dish plan was initially determined.

The preliminary plan was finalized and the practice began on the second floor of Quanjude Roast Duck Restaurant at Hepingmen, Beijing, where a large exhibition stand of 16 square meters was set up in the spacious hall. Since the spring of 1985, Li Qigui, Chen Shoubin and Zhang Zhiguang had been practicing there, striving to be confident and adept. For several months, they had tried a lot of dishes, some of which were not very successful, some not suitable for live performances, some spent too long to operate and thus not suitable for the rhythm of the competition, so they tried and adjusted, constantly eliminating the dishes and updating the dishes. For seven or eight months, they, like scientists in the laboratory, had been experimenting repeatedly and making unremitting efforts for the first appearance of New China's cooking skills on the international stage.

Their final plan was to make a large-scale food carving work Heavenly Maids Scattering Blossoms in the middle of the 16-square-meter large exhibition stand, surrounded by assorted cold dishes Magpie on Plum Tree and Dragon and Phoenix. The eight hot dishes included Baoshen & Wax Gourd Cup, Beijing Roast Duck, Hot Pot with Eight Treasures, Flowers & Pearl Fish, Litchi & Duck Roll, Sauteed Prawns with Hericium Erinaceus, Top-graded Lotus and Father-Son Lobster. There were two dishes sold on the spot: Sweet and Sour Fish Sticks and Crispy Duck Legs.

Because they went to Europe for the competition, they must understand some local customs and etiquette, and also master some eating habits. For example, when you went to Europe, you should get used to drinking ice water, using knives and forks, and eating steak. It's also a required course for Li

于李启贵他们也是一门必修课。于是他们被安排到北京的马克西姆餐厅学习西餐礼仪，李启贵也请宋怀桂等马克西姆餐厅的朋友们，到泰丰楼品尝他的拿手鲁菜。

光阴似箭，日月如梭，转眼到了参赛赴欧的日子。1985年12月25日，正赶上西方的圣诞节，中国烹饪代表团的飞机降落在卢森堡机场。圣诞节期间的卢森堡装扮得如同童话世界一般，33岁的李启贵第一次走出国门，眼前的一切让他倍感惊奇。

他们一行五人顾不上休息，也顾不上逛街看景，而是带着25箱参赛的炊具、白酒、八仙人的花雕酒，先到组委会接待处报到寄存东西去了。

展台布置好，万众聚焦的第五届奥林匹克世界烹饪大赛旋即开始了。卢森堡不大的山城里，聚集了数以万计的热情观众，他们都是闻讯从世界各地赶到这里，为的就是见证全球厨神当场献艺的精彩瞬间。每个干净开阔的大展台，都设在一个大玻璃房子里，来自77个国家和地区的厨师们各显神通，各展绝技。一时间，整个比赛现场香气四溢，锦簇花团，雪亮的厨刀在砧板上欢快地跳舞，食品搅拌机在一旁铿锵地歌唱。观众和评委们把感官尽情打开，乐享着现场最奇妙最美好的饮食文化。这其中，由于中国代表团是第一次在本项赛事上出场亮相，左边挨着苏联代表团，右边挨着美国代表团，俄式大餐、美式西餐、中华厨艺，几个大国之间在烹饪上同台竞技，所以备受瞩目。

展台上，李启贵与张志广合作，把事先刻好的四大盆"菊花""大丽花""月季"等五彩食材，精心地摆出"天女散花"的造型，展台上"荷花"盛开，八只冬青草雕成的小仙鹤向展台的四方振翅欲飞。两侧由香菇丝、各色肉丝、蔬菜丝组成的"喜鹊登梅""龙凤呈祥"两个冷拼盘也布置完成，展台前一下子拥过来无数镜头和好奇的目光。

选手进了场，组委会工作人员才根据选手提交的供料单提供比赛原料。不成想就在这个环节上第一个麻烦来了。仿佛是命运跟李启贵开的玩笑，现场销售的"糖醋鱼条"在原料上出了岔子：组委会给李启贵提供的主料，比目鱼个头太小了！本来大比目鱼肉厚、质嫩、刺又少，改刀方便，烹饪简单，酸甜口味儿在适合西方观众口味的同时，还有新鲜感。结果面对着一堆不够材料的小比目鱼，李启贵摇了摇头，二话没说，提刀上阵。尽管李启贵是出了名的快手儿师傅，收拾出这130份糖醋鱼条的原料，也着实考验着他的基本功和耐力。在李启贵的身上，总有一股子不服输不怯阵的劲头儿。只见他手中的刀上下翻

Qigui and his team. So they were arranged to study Western food etiquette at Maxim's in Beijing. Li Qigui also invited friends from Maxim's such as Song Huaigui to taste his specialties of Shandong cuisine at Taifeng Restaurant.

Time flies like a shuttle. It's time to go to Europe for the competition soon. On December 25, 1985, just on the occasion of Christmas in the West, the plane of the Chinese cooking delegation landed at Luxembourg Airport. Luxembourg was dressed up like a fairy tale world during Christmas. Li Qigui, at the age of 33, went abroad for the first time and was surprised at everything he saw.

The delegation of five chefs hurriedly took 25 boxes of cooking utensils, liquor and high-grade Shaoxing wine named Shaobaxian to the reception desk of the Organizing Committee and checked in.

The stand was well arranged, and the 5th IKA that all eyes focused immediately began. In Luxembourg, a small mountain city, tens of thousands of enthusiastic spectators gathered together. They all arrived there from all over the world to witness the wonderful moment when the global top chefs performed on the spot. Each clean and open large exhibition stand was stood in a large glass house. Chefs from 77 countries and regions showed their unique skills. For a moment, the whole competition scene was full of aroma of dishes, with clusters of flowers brocading, bright kitchen knives dancing happily on the chopping board, and the food mixer singing sonorously. The spectators and the judges enjoyed the most wonderful and beautiful food culture on the scene with open mind. Among them, as the Chinese delegation attended this competition for the first time, with the Soviet delegation on the left and the U.S. delegation on the right, i.e., Russian cuisine, American western food and Chinese cooking. Several powers competed on the same stage in cooking, so it had attracted much attention.

On the stand, Li Qigui cooperated with Zhang Zhiguang to carefully pose the four large pots of colorful ingredients such as chrysanthemum, dahlia and rose carved in advance as the shape of Heavenly Maids Scattering Blossoms. The "lotus" on the stand was in full bloom, and eight small cranes carved from winter grass fluttered their wings to the four sides of the stand. On both sides were the two assorted cold dishes Magpie on the Plum and Dragon and Phoenix, which were composed of shredded mushrooms, shredded meat of various colors and shredded vegetables. Numerous scenes and curious eyes suddenly crowded in front of the exhibition stand.

Only when the contestants entered the competition did the staff of the Organizing Committee provide the competition materials according to the materials list submitted by the contestants. Unexpectedly, the first trouble came in this link, as if it were a joke that God played on Li Qigui. Sweet and Sour Fish Sticks sold on the spot had some trouble with the main ingredients: the flounders, as the main ingredient provided by the Organizing Committee to Li Qigui, were too small! Originally, it's thought that flounder was thick in meat, tender in quality and few in thorns, so it's convenient to cut and easy to cook and the sweet and sour taste was not only suitable for Western audiences, but also fresh. However, facing a group of small flounders with insufficient materials, Li Qigui shook his head and started work without any words. Although Li Qigui was a well-known fast-cutting master, picking up the main ingredients for 130 pieces of Sweet and Sour Fish Sticks also really tested his basic skills and endurance. In Li Qigui's mind, there was always an indomitable spirit. With the knife in his hand flying up and down, snow-white fish came in a great deal. The three sharpened knives

飞，雪白的鱼肉纷披而至，磨好的三把刀都用上了，去鱼皮、剔除鱼刺、改刀鱼条，李启贵出了几身透汗，一场有惊无险的考验过关了。随后制作的香酥鸭腿，李启贵反复刀工处理，调料腌制，更是驾轻就熟，顺利完成。

每道现场销售菜完成后，还要配上相应的主食，比如米饭或是馒头。没想到第二个麻烦跟着就来了：组委会没准备蒸锅！

老话说，不蒸馒头争口气，没有蒸锅，别说为国争光，蒸啥也蒸不了啦！眼看着比赛时间一分一秒地过去，组委会的工作人员一边解释表示歉意，说中国饮食文化历史悠久，这又是你们第一次参赛，我们在中餐厨具方面的确准备不足，一边不停地拧着眉毛耸着肩，表示实在无能为力。李启贵的脑门上冒出了一层汗珠子。怎么办？说好的一鸣惊人呢？说好的勇夺金牌呢？没有蒸锅，准备不出配餐的主食，就意味着没有完成比赛的规定动作，等于前功尽弃，淘汰出局。陪同的使馆工作人员表示，使馆里倒是有一口蒸锅，可是鞭长莫及，往返取锅的时间来不及了。

莫非希望即将成为泡影？李启贵怀里如同揣着25只小老鼠，百爪挠心。就在这一筹莫展的当口，绝处逢生。展台不远处，忽然挤过来一位五十来岁的华人厨师，右手高高举着一口蒸锅，左手像击鼓一样，把那口锅"啪啪"拍得山响："别着急！我这儿有，你们先用！"这真是山重水复疑无路，柳暗花明又一村！喜出望外的李启贵心里觉得热乎乎的，连忙接过老先生手中的蒸锅，道了一声谢，就继续忙着参赛，后来还锅的时候，竟然忘了请教人家的尊姓大名，这让李启贵懊恼不已。他一直打听这位热心华侨的下落，直到2007年，也就是22年后，他才通过蒋经国先生的大厨王清标先生得知，当年雪中送炭给他们送来蒸锅的老华侨，就是新加坡前总理李光耀的娘惹菜大师叶高和先生。再见面时，叶大师已经七十多岁了，腿脚有些不便，须拄杖前行。叶先生说，那个时候在海外看到中国人就特别亲，愿意尽己所能地帮助你们。你们果然为祖国拿了金牌争了光！这是后话。

回到1985年的比赛现场，李启贵制作的各130份糖醋鱼条、香酥鸭腿是77个代表团中第一个销售完成的，受欢迎程度由此可见。

随后李启贵大师又连续完成了宝参冬瓜盅、八宝涮锅、百花珍珠鱼、荔枝鸭卷、猴头菇燔大虾、一品芙蓉、父子龙虾等七道热菜的烹制装盘。与此同时，在现场没有烤炉的情况下，陈守斌大师也完成了北京烤鸭的制作。

were all used, removing the fish skins, fishbone, and cutting the fish into fish sticks. Li Qigui was sweating for several times and passed a near-miss test. The crispy duck legs made later were treated by Li Qigui repeatedly with knives and pickled with seasonings, which he was even more adept at and successfully completed.

After each dish sold on the spot was completed, the corresponding staple food should be provided, such as rice or steamed bread. It's unexpected that the second trouble arised: The Organizing Committee didn't prepare steamers!

As the old saying goes in Chinese, people must try to make a good showing. If you don't have a steamer, you couldn't steam anything, let alone win honor for the China. The clock ticked away. The staff of the Organizing Committee made apologies and explained that Chinese food culture had a long history and that was the first time for Chinese chefs to participate in the competition so they were not well-prepared for Chinese food and kitchenware, expressing they were helpless. Li Qigui's forehead was full of sweat beads. How to do? What about the blockbuster? What about winning the gold medal? No steamer and staple food meant that the required actions of the competition had not been completed, which was tantamount to losing all previous achievements and being eliminated. Accompanied embassy staff said that there was a steamer in the embassy, but it's too late to get the steamer back and forth since it's too far.

Did hope come to naught? Li Qigui was like ants on the hot pot. At this juncture of losing ideas, they found rescue in desperate circumstances. Not far from the stand, suddenly a Chinese chef at his fifties came over through the crowd, holding a steamer high in his right hand and beating it as a drum with his left hand, "Don't worry! I have it here, you use it first!" This was really snatched from the jaws of death! Overjoyed, Li Qigui felt warm in his heart. He hurriedly took the steamer in the old man's hand, thanked him and continued to participate in the competition. Later, when he returned the steamer, he forgot to ask his name. This made Li Qigui annoyed so that he had been inquiring about the whereabouts of this enthusiastic overseas Chinese. It was not until 2007, that's, 22 years later, that he learned through Mr. Wang Qingbiao, Mr. Jiang Jingguo's chef, that the old overseas Chinese who gave them timely assistance was Mr. Ye Gaohe, the master of Nyonya for former Singapore Prime Minister Lee Kuan Yew. When they met again, Master Ye was already in his seventies and with some trouble in feet and legs, he had to walk on a stick. Mr. Ye said that when he saw the Chinese overseas, he was very pleased and willing to do his best to help them. They did win the gold medal for the motherland! This was another story.

Recalling the competition site in 1985, 130 parts of Sweet and Sour Fish Sticks and Crispy Duck Legs made by Li Qigui were sold up in the first time among 77 delegations, thus showing the popularity.

Later, Master Li Qigui successively cooked and dished out seven hot dishes, including Baoshen & Wax Gourd Cup, Hot Pot with Eight Treasures, Flowers & Pearl Fish, Litchi & Duck Roll, Sauteed Prawns with Hericium Erinaceus, Top-graded Lotus and Father-Son Lobster. At the same time, in the absence of an oven at the scene, Master Chen Shoubin also completed the production of Beijing Roast Duck.

据李启贵大师回忆，由于中餐销售现场气氛过于热烈，购餐的客人还排着长长的队伍，世界烹饪联合会的主席和秘书长临时决定，大赛的开幕式推迟40分钟举行。当他们售卖完进入赛场的时候，全场迸发出热烈的掌声，欢迎中国队正式入场。李启贵感到无上的荣光，也更加真切地感受到中国博大精深的饮食文化，在世人面前展现出的巨大魅力。这场面是奥林匹克世界烹饪大赛开赛以来从没有出现过的情景，也深深地印在了李启贵的脑海中。

比赛的结果当场揭晓，以李启贵、陈守斌、张志广为主力选手的中国烹饪代表团，获得了组委会的金质奖牌，称得上实至名归。汗水没有白流，付出终有回报，最让李启贵觉得有成就感的，就是他凭借高超的烹调技艺和良好的团结协作精神，为新中国赢得了世界烹饪顶级赛事的第一块金牌。他甚至觉得，这和1984年许海峰在洛杉矶奥运会上夺得的那枚射击金牌的分量差不多。

为了给中国代表团庆功，更是为了以饮食文化增进中卢两国之间的友谊。中国驻卢森堡大使特地在大使馆设首相宴，邀请卢森堡大公国的首相夫妇莅临宴会，品尝中华美食。自然，主厨的李启贵、陈守斌再次大显身手。当晚的菜品包括一品芙蓉、宝参冬瓜盅、猴头菇燸大虾、百花珍珠鱼、北京烤鸭、鲍鱼菜心、清烹牛柳丝、龙须扒菜心等，首相先生赞不绝口，为自己能够吃到最正宗的中国佳肴感到十分荣幸。其中李启贵制作的八生火锅最受首相先生青睐。火锅是景泰蓝的工艺，造型古朴庄重，锅围上是龙凤呈祥的纹饰，宝蓝明金，熠熠生辉，再配上李启贵大师制作的八生菜肴，可谓"色香味形器"五美俱全。宴会结束时，中方把这只精美的八生火锅赠送给了首相先生。

忙完大使馆的首相宴，李启贵大师没有闲下来。他被邀请到卢森堡国际俱乐部授课，为二十多国的观众和厨师讲授了中国烹饪理论和中国参赛的名菜，让听众们感到耳目一新，受益匪浅。访欧期间，代表团还考察了法国、德国、比利时、卢森堡等国的餐饮业态。其中最值得一提的是，李启贵在卢森堡帝国饭店交流表演时的奇遇。

在帝国饭店的大餐厅里，李启贵边讲边做，锅烧牛肉、糖醋鱼条、香酥鸭腿、香辣青椒牛柳丝、干燸大虾、白扒猴头蘑、油焖大虾、龙须菜心、清

As Master Li Qigui recalled, the Chinese food was so popular on the spot with long queue of guests for buying, the president and secretary-general of the World Cooking Federation decided temporarily to postpone the opening ceremony of the competition by 40 minutes. When they entered the stadium after selling, the whole stadium burst into warm applause to welcome the Chinese team into the stadium. Li Qigui felt great glory and more closely felt the great charm of Chinese extensive and profound food culture in front of the world. This scene had never appeared since the start of the IKA, and was deeply imprinted in Li Qigui's mind.

The result of the competition was announced on the spot. The Chinese cooking delegation, with Li Qigui, Chen Shoubin and Zhang Zhiguang as the main players, won the gold medal from the Organizing Committee, which was truly worthy of the name. All efforts eventually deserved. What made Li Qigui feel most fulfilled was that he won the first gold medal in the world's top cooking competition for New China with his superb cooking skills and good spirit of unity and cooperation. He even felt that this was similar to the gold medal won by Xu Haifeng in shooting at the Los Angeles Olympics in 1984.

In order to celebrate the Chinese delegation as well as to promote the friendship between China and Luxembourg through food culture, the Chinese Ambassador in Luxembourg specially held a banquet for the prime minister at the embassy, inviting the prime minister and his wife from the Grand Duchy of Luxembourg to the banquet and taste Chinese dishes. Naturally, chefs Li Qigui and Chen Shoubin once again showed their talents. The dishes of the banquet included Top-graded Lotus, Baoshen & Wax Gourd Cup, Sauteed Prawns with Hericium Erinaceus, Flowers & Pearl Fish, Beijing Roast Duck, Cabbage and Abalone, Stir-fried Shredded Beef, Sauteed Green Vegetable with Asparagus, etc. The Prime Minister praised profusely and felt very honored to be able to taste the most authentic Chinese delicacies. Among them, the hot pot with eight fresh ingredients cooked by Li Qigui was most popular with the Prime Minister. The hot pot was cloisonne, with a simple and solemn shape. The pot enclosure was decorated with dragon and phoenix, with sapphire blue and bright gold, and shined brightly. When matched with the hot pot with eight fresh ingredients cooked by Master Li Qigui, it could be said that it was completed in "five beauties such as color, aroma, taste, shape and utensil". At the end of the banquet, the Chinese side presented this exquisite hot pot to the Prime Minister.

After the banquet for the prime minister at the embassy, Master Li Qigui had no spare time. He was invited to give lectures at Luxembourg International Club, presenting audiences and chefs from more than 20 countries about Chinese cooking theory and famous Chinese dishes. The spectators felt refreshing and benefited greatly. During their visit to Europe, the delegation also inspected the catering industry in France, Germany, Belgium, Luxembourg and other countries. Among them, the most noteworthy was Li Qigui's adventure during the exchange performance at the Hotel Empire in Luxembourg.

蒸活鱼、蒜蓉蒸龙虾、牛柳粒生菜包……一道道活色生香的中式菜肴，受到现场各界人士的热情赞许。在表演结束的时候，帝国饭店的老板忽然找到李启贵说："李，我认识一位法国朋友，他知道你刚刚获得了金牌，想跟你学习厨艺，可以吗？"李启贵爽快地答应了对方的请求。很快，这位法国小伙子带着翻译就来到李启贵面前。李启贵说："既然你想学，我就教你四道中国菜的做法，就是蒜蓉蒸龙虾、清烹牛柳丝、牛肉粒生菜包、油焖大虾。"

一上手李启贵发现，小伙子不是一时兴起，而是真学。这虾怎么开？龙虾怎么修理？牛肉怎么去筋？洋葱、牛肉怎么切成粒？怎么煸怎么炒？李启贵一一地从主料、配料、调料、制作方法、口味炒给他看，手把手教给他。这小伙子学得非常认真。李启贵好奇地问他为什么要学做中式菜肴？他说他是做法餐的，但是他要学习了解中国悠久的饮食文化，要跟中国大厨学几样地道的中国菜。

过了几天，中国代表团的表演活动结束，在他们即将离开的时候，这个法国洋学生又来找李启贵了，热情地邀请他的李启贵老师和其他几位中国代表团成员到他的店里去看一看。代表团跟大使馆汇报了一下，就应约去了。到了那儿才知道，原来这个学中餐的法国小伙子是一家高档法式大酒店的老板，特地请他的中国老师和朋友们来品尝法式大餐。一时间，牛排、烤鱼、煎鹅肝、法式焗蜗牛、蔬菜沙拉，全都端上来了。

厨师的眼光和寻常客人就是不一样，李启贵大师发现，这里的食材绝对是上乘纯精选的，不仅牛排的鲜嫩程度给人带来口感上的愉悦，而且他们使用的厨房机械自动化设备也是在世界上领先的。这让李启贵心生艳羡：咱们中国什么时候也能用上这样先进的炊事设备就好啦！

代表团回到北京已经是1986年1月底了。北京市有关部门给载誉而归的中国烹饪代表团隆重颁奖。奖励了每位成员一台录像机，由于李启贵大师贡献尤其突出，奖励他的是一辆本田125两轮摩托车。令许多人奇怪的是，李启贵声名鹊起之后，却在人们的视野中忽然消失了。他到底去哪儿了呢？

In the grand hall of the Hotel Empire, Li Qigui was teaching while cooking Roasted Beef in Pot, Sweet and Sour Fish Sticks, Crispy Duck Legs, Spicy Shredded Beef with Green Pepper, Dried Simmered Shrimp, Steamed Hericium Erinaceus, Braised Shrimp in Oil, Asparagus and Cabbage, Steamed Live Fish, Steamed Lobster with Garlic, Beef Fillet Wrapped with Lettuce ... Chinese dishes with live color and fragrance received warm praises from all walks of life on the scene. At the end of the performance, the boss of the Hotel Empire suddenly found Li Qigui and said, "Li, I know a French friend. He knows that you have just won the gold medal and wants to learn cooking from you, OK?" Li Qigui nicely agreed to his request. Soon, the French young man came to Li Qigui with an interpreter. Li Qigui said, "Since you want to learn, I will teach you four Chinese dishes, namely Steamed Lobster with Garlic, Stir-fried Shredded Beef, Beef Fillet Wrapped with Lettuce, and Braised Prawns with Oil."

Li Qigui found that the young man was not on a whim, but a real student. How to open this shrimp? How to cut lobster? How to remove tendons from beef? How to cut scallions and beef into grains? How to stir-fry and how to fry? Li Qigui showed him the main ingredients, ingredients, seasonings, making methods and tastes one by one, and taught him hand in hand. The young man studied very hard. Li Qigui asked him curiously why he wanted to learn cooking Chinese dishes. He said that he was devoted to cooking French dishes, but he wanted to learn about China's long-standing food culture and learnt several authentic Chinese dishes from Chinese chefs.

A few days later, the Chinese delegation ended the performance. Just as they were about to leave, the French foreign student came to Li Qigui again and warmly invited his teacher Li Qigui and several other members of the Chinese delegation to visit his restaurant. The delegation reported to the embassy and went on an appointment. Only when they got there did they know that this French young man who studied Chinese food was the owner of a high-end French restaurant and specially invited his Chinese teachers and friends to taste French dishes. After a short while, steak, grilled fish, fried foie gras, French baked snail and vegetable salad were all served.

The chefs' vision was different from that of ordinary guests. Master Li Qigui found that the ingredients here are absolutely excellent and purely selected. Not only did the freshness and tenderness of steak bring pleasure to people in taste, but also the mechanical automation equipment they used in the kitchen was the world's leading ones. This made Li Qigui envious: When could we use such advanced cooking equipment in China?

The delegation returned to Beijing at the end of January 1986. Beijing authorities gave a grand award to the Chinese cooking delegation that returned with honor. Each member was rewarded with a video recorder. Thanks to the outstanding contribution of Master Li Qigui, he was rewarded with a Honda 125 two-wheeled motorcycle. To the surprise of many people, after Li Qigui became famous, he suddenly disappeared from people's view. Where the hell had he gone?

1986年，李启贵大师从卢森堡"奥林匹克世界烹饪大赛"上载誉归来，获得嘉奖。
In 1986, Li Qigui won a prestigious award in Luxembourg for his participation in the IKA/Culinary Olympics in Luxembourg.

1986年，李启贵大师参加"奥林匹克第五届世界烹饪大赛"，荣获中国首块金牌。
In 1986, Li Qigui won China's first gold medal at the 5th IKA/Culinary Olympics.

1986年，李启贵大师参加"奥林匹克第五届世界烹饪大赛"，颁奖现场。
In 1986, Li Qigui participated in the 5th IKA/Culinary Olympics and was awarded a prize.

李启贵大师在欧洲卢森堡表演中国名菜。

Li Qigui performed the cooking of famous Chinese dishes in Luxembourg, Europe.

1986年，李启贵大师参加"奥林匹克第五届世界烹饪大赛"荣获金牌，回国后北京市副市长郭献瑞在庆功会上为李启贵大师颁奖。

In 1986, Li Qigui won the gold medal at the IKA/Culinary Olympics. After he returned to China, Guo Xianrui, Vice Mayor of Beijing, presented an award to Li Qigui at the celebratory ceremony.

2000年1月28日，李启贵大师烹制贺岁名菜"风生水起"，刊登在《北京日报》上。

On January 28, 2000, Li Qigui cooked the famous dish for the Chinese New Year—Fengsheng Shuiqi and was covered by Beijing Daily.

2000年2月13日，李启贵大师出席由中央电视台举办、侯耀文主持的烹饪节目"全家福"，中国大陆、香港、澳门、台湾的四位大师烹制名菜，共庆新春。

On February 13, 2000, Li Qigui attended the Chinese, Hong Kong, Macau and Taiwan culinary program "Family Portrait" hosted by Hou Yaowen at CCTV, where four masters cooked famous dishes to celebrate the Chinese New Year.

第四回

Chapter Four

进退有据 重金不卖一招鲜
大师驾到 京派鲁菜迷港岛

李启贵从卢森堡拿回中国第一块国际烹饪大赛金牌之后，并没有躺在功劳簿上睡大觉。相反，他通过在欧洲的考察和交流，想到了更多的东西。在国外，比如客人在中餐厅吃京派鲁菜，忽然提出来，既然是京派鲁菜，你这儿有没有王府菜？官府菜？宫廷菜？人家也可能要点一些中国其他七大菜系的菜看，厨师会不会被客人"考住"呢？有道是"艺不压身"，而且师父王义均大师也一直鼓励他要多学多看，转益多师，千万别取得点儿成绩就沾沾自喜，故步自封。李启贵对师父开明豁达的胸怀更加崇敬，他决定把自己"藏起来"，到西绒线胡同著名的四川饭店学做川菜。

拿定主意，李启贵跟单位领导打好招呼，说为了厨艺更加全面，自己请求去四川饭店深造川菜。接着，他通过朋友找到四川饭店的经理侯兴林表明了自己的来意。侯经理一听，有点儿犯难："您是国际金牌的获得者，又是泰丰楼的厨师长，身份这么特殊，到了后厨我怎么介绍你呀！"李启贵呵呵一笑说："这好办，干脆您就说，我是做机关餐的，食堂炒大锅菜的大师傅来实习的。"

侯兴林经理一看，也只好如此，就替李启贵隐瞒了真实的身份。李启贵在四川饭店的后厨一干就是小一年，跟着四川籍的厨师长陈松如老师傅学做川菜，当时还有四川的唐文安师傅、刘子华师傅和北京的郑绍武师傅。郑绍武后来和李启贵先后拜王义均大师为师，成了师兄弟，这是后话。

在后厨工作的间隙，几位川菜师傅总忍不住打量眼前这位浓眉大眼、身材魁伟的"实习生"。忽然有人说："咦，看你很眼熟啊！"李启贵赶紧扭着脸儿说："不可能啊。""不对，好像在报纸上、电视上见过你。"另一位师傅在旁边一拍大腿："你是不是那个出国拿大奖的呀？"眼看着要露馅儿，李启贵倒是一脸镇静，他掸了掸身上的面粉说："此言差矣！天下衣帽相形的人物甚多，很可能是你看错了。"就这样，李启贵使了这招儿叫"白衣渡江——蒙混过关"，居然对付过去了。直到1987年，李启贵觉得学有所成，单位那边又一个劲儿地催他回泰丰楼管事，这才结束了隐姓埋名的偷艺生涯，川菜手艺已经学到了身上。

Adhering to the Principle and Refusing to Sell Cooking Skills Even at A Huge Sum of Money
Beijing-style Shandong Cuisine Attracting Great Attention in Hong Kong

After taking back China's first gold medal in the IKA from Luxembourg, Li Qigui did not stop working with his credit. On the contrary, he thought of more things through his investigation and exchange in Europe. In foreign countries, for example, when guests ate Beijing-style Shandong cuisine in Chinese restaurants, they suddenly asked, "Since it is Beijing-style Shandong cuisine, do you have palace cuisine here? Government cuisine? Royal cuisine?" They would also order some dishes of the other seven major Chinese cuisines. Would the chefs be "tested" by the guests? There was a saying that "Knowledge is no burden", and Master Wang Yijun had always encouraged him to learn more and watch more, learning from more masters, and not to be complacent just because he had made some achievements. Li Qigui was even more reverent for Master's enlightened and open-minded mind. He decided to "hide" himself and learn to cook Sichuan cuisine at the famous Sichuan Hotel in Xirongxian Lane.

Making up his mind, Li Qigui reported the leaders of the unit and said that in order to cook more comprehensively, he requested to go to Sichuan Hotel to further study Sichuan cuisine. Then, he found Hou Xinglin, Manager of Sichuan Hotel, through his friends and expressed his intention. Hearing this, Manager Hou was a little puzzled, "You are the winner of the IKA and the head chef of Taifeng Restaurant. You have such a special status. How should I introduce you to the back kitchen?" Li Qigui said with a smile, "It's easy. Just say that I am a chef cooking office meals and dishes of large pots in the canteen and come here for internship."

Manager Hou Xinglin had to do so without any other options. He hid the true identity of Li Qigui. Li Qigui worked in the back kitchen of Sichuan Hotel for a year. He learned to cook Sichuan cuisine with Sichuan Master Chen Songru, who was also a head chef. At that time, there were also Sichuan chefs such as Tang Wenan and Liu Zihua and Beijing chef Zheng Shaowu. Zheng Shaowu and Li Qigui later took Master Wang Yijun as their instructor and became fellow apprentices. This was another story.

During the breaks while he worked in the back kitchen, several Sichuan cuisine chefs couldn't help looking straight at the heavy-eyed and tall "intern" in front of them. Suddenly someone said, "Well, you look familiar!" Li Qigui quickly turned his face and said, "Impossible." "No, it seems to have seen you in the newspaper or on TV." Another master slapped his thigh nearby, "Are you the one who won the grand prize abroad?" It's going to let the cat out of the bag. Looking calm, Li Qigui dusted his flour and said, "It's not true! There are many people in the world who look like in clothes and hats. It is very likely that you are wrong." In this way, Li Qigui slipped through in the end. It was not until 1987 that Li Qigui felt that he had achieved something in his studies, and the unit kept urging him to return to Taifeng Restaurant to take charge. Only then did he end his career of stealing cooking skills under anonymity, and he had learned Sichuan cuisine skills then.

1989年午暖还寒时候，一家日本高级酒店的老板加藤靖三先生从山东旅游来到北京。在游览了天安门和广场之后，来到了前门东侧的泰丰楼饭庄。他们在楼上的包间里点了一桌菜，有鱼有虾有牛肉有青菜，越吃越高兴。随即他们通过服务员找到泰丰楼的经理李继和，说想见见这个做菜的厨师，并且表示能否以公派的方式，介绍这个厨师到他的酒店去工作一段时间。过了几天，又拜托朋友来沟通洽谈。经过反复磋商，就把李启贵和他的一个助手请到了日本大阪的南海假日酒店。来日本之前，好学的李启贵专门在北京报了班，学了半年的日语。有准备的人生，看上去总是那么与众不同。

南海假日酒店是大阪的一家五星级高档酒店，里面设有多家风味餐厅。比如花香楼是日餐厅，桃源是中餐厅，还有法式餐厅、酒吧及多功能厅等。李启贵供职的餐厅就是桃源。当时大阪有五千多家餐饮店，但是真正经营中国菜的不算多，而且其中不少已经中餐日本化了。李启贵的到来，让几十种正宗的山东名菜闪亮登场，顿使大阪的中餐面貌为之一振。

加藤先生是个中国迷，更迷李启贵的京派鲁菜。几乎每天中午他都到桃源餐厅吃饭，就为了品尝李启贵的手艺。李启贵则为了弘扬中华优秀传统饮食文化，尽展才华，天天换新菜，几个月不带重样儿的。这可把包括加藤在内的所有日本同事镇住了，大家不知道李启贵脑子里的菜谱到底有多长？

当然，并不是所有日本人都对中国人特别友好，也有戴着有色眼镜看待李启贵的。似乎觉得李启贵之所以如此风光，不过是老板一时图新鲜，本身并没有什么真本事。然而事实胜于雄辩，一个偶然的事件或许就能改变人们先入为主的看法。一天，加藤先生有一个重要宴会，他请李启贵大师提前两天准备材料，足见非同寻常。

当天的宴会进行到高潮时，李启贵大师在现场表演了拿手绝活——抻龙须面。平展展的大案板上，一大堆雪花粉纷纷扬扬，在李启贵的双掌之间起起落落，与清水深度融合，渐渐地，"雪花"变成了肌肤般洁净的面团，润泽而富有弹性。接着，李启贵反复遛面，使其更具韧性和活力。包括日本官房长官在内的满场用餐的贵宾们，都不由自主地停下了杯箸，目不转睛地看着李启贵大师的精彩表演，连后厨的日本厨师也暂时放下手里的活儿，站在一旁"偷艺"。李启贵身材高大魁梧，双臂展开，宽有丈余。沉甸甸的面团在他手里被轻松地抻来勾去，随意化形，整整13扣，细如发丝的龙须面被抻成两米多长，用手指挑起来向众人展示，竟然根根不断，如同一锭抖开的白纱线！顿时举座皆惊。

At the beginning of spring in 1989, Mr. Shinzo Katô, owner of a senior Japanese hotel, came to Beijing from Shandong. After visiting Tiananmen Square, he came to Taifeng Restaurant on the east side of Qianmen with his delegation. They ordered a table of dishes in the private room upstairs, with fish, shrimp, beef and green vegetables, which they enjoyed greatly. Then they found Li Jihe, Manager of Taifeng Restaurant, via the waiter, and said that they wanted to meet the cook and whether he could introduce the cook to work for a period of time in his hotel in a manner of public dispatch. After a few days, they sent their friends to communicate and negotiate. After repeated discussions, Li Qigui and one of his assistants were invited to the Holiday Inn Nankai in Osaka. Before coming to Japan, Li Qigui studiously enrolled in a special class in Beijing and studied Japanese for half a year. A prepared life always looks so different.

Holiday Inn Nankai was a five-star high-end hotel in Osaka, with many flavor restaurants provided. For example, Huaxiang Restaurant was a Japanese restaurant, Taoyuan Restaurant was a Chinese restaurant, and there were French restaurant, bar and multi-function hall. Li Qigui worked in Taoyuan Restaurant. At that time, there were more than 5,000 catering companies in Osaka, but few of them actually operated Chinese food, while many became Japanese-style Chinese food. With the arrival of Li Qigui, dozens of authentic famous Shandong dishes had come on stage, giving a boost to Osaka's Chinese food.

Mr. Katô was a fan of China as well as a fan of Li Qigui's Beijing-style Shandong cuisine. Almost every noon, he went to Taoyuan Restaurant for dinner just to taste dishes cooked by Li Qigui. Li Qigui, on the other hand, in order to carry forward the excellent traditional Chinese food culture and display his talents, he cooked new dishes every day without repetition for several months, which confused all Japanese colleagues, including Katô, and they didn't know how many dishes Li Qigui could cook.

Of course, not all Japanese were particularly friendly to Chinese, and some had a strong bias against Li Qigui. It seemed that Li Qigui was so popular just because the boss liked Chinese dishes temporarily for curiosity and he didn't have any true skills. However, facts speak louder than words, and an accidental event may change people's preconceived views. One day, Mr. Katô had an important banquet. He asked Master Li Qigui to prepare materials two days in advance, which indicated that the banquet was unusual.

At the climax of the banquet on that day, Master Li Qigui performed his unique skill of pulling Longxu Noodles on the spot. On the flat large chopping board, Li Qigui blended a plenty of snowflake-like flour with clear water, which gradually became skin-like clean dough, moist and elastic. Then, Li Qigui kneaded dough repeatedly to make it more resilient and vibrant. The distinguished guests, including Japan's chief cabinet secretary, all stopped their chopsticks involuntarily and watched Master Li Qigui's wonderful performance intently. Even the Japanese chef in the back kitchen temporarily stopped his work and stood aside to "steal skills". Li Qigui was tall and burly, and outspread arms had a width of more than ten feet. The heavy dough was easily pulled around in his hands and randomly shaped. With 13 folds, the hair-thin Longxu Noodles were pulled into more than two meters long. He picked up with his fingers to show them to the public. They were just like a spindle of white yarn shaking away, immediately surprised all.

 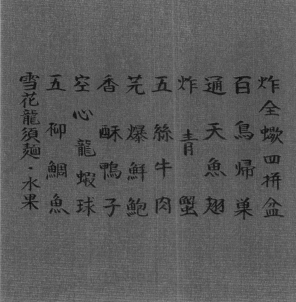

1991年12月1日至1992年1月31日，李启贵大师在日本烹制的北京名菜菜单。
From December 1, 1991 to January 31, 1992, menu of famous Beijing dishes prepared by Li Qigui in Japan.

1990年，李启贵大师在日本大阪五星级酒店工作期间迎接了北京市饮食服务总公司派往日本考察的师父王义均。
In 1990, during his work in a five-star hotel in Osaka, Japan, Li Qigui met his mentor Wang Yijun, who was sent to Japan by the Beijing Catering Service Corporation for a study tour.

1992年4月，李启贵大师在日本大阪和日方厨师长片町先生赏樱花。
In April 1992, Li Qigui admired cherry blossoms in Osaka, Japan with Mr. Katamachi, the head chef of the Japanese side.

2000年，第三届中国烹饪世界大赛在日本东京新高伦王子饭店举行。
In 2000, the 3rd World Competition of Chinese Cuisine was held at the Grand Prince Hotel Shin Takanawa in Tokyo, Japan.

2000年，在日本的新高伦王子饭店，与第三届中国烹饪世界大赛的获奖选手合影。
In 2000, Li Qigui posed with the winners of the 3rd World Competition of Chinese Cuisine at the Grand Prince Hotel Shin Takanawa in Japan.

这一天，李启贵到大阪工作还不到一个月。中日两国厨师间的友好局面也在这一刻打开了。

抻龙须面的事儿没过多久，李启贵又在日本厨师面前露了一手绝活。桃源餐厅的中餐宴会档次比较高，李启贵就别出心裁地推出了一道新菜品"空心龙虾球"。

本来提起制作海鲜菜品，日本厨师多少有点儿自诩。日本是海中岛国，靠海吃海，海鲜刺身那是日本料理的看家本领。而来自北京泰丰楼的中餐大厨李启贵的海鲜手艺如何，他们在心里还是打了个问号。当然，这个"问号"不好直接说出来，于是加藤先生让人从海鲜市场买回一只六斤重的大龙虾。他拎着长长的虾须子走进厨房，笑呵呵地对李启贵说："李桑，你晚上吃这个。"李启贵高高兴兴地接过来，心说晚上老板请我吃这个？太破费了！可转念一想，不对劲儿，这里边有套头儿！他是想让日本厨师们看看我会不会处理龙虾，有没有做海鲜的手艺。

想到这儿，李启贵轻松一笑。从备战奥林匹克世界烹饪大赛开始，师父王义均就把海鲜的制作方法都传授给他了。对于曾经在世界大赛上夺冠的人来说，眼前这个"龙虾试卷"简直是"张飞炒豆芽——小菜一碟"。当天晚上，李启贵把龙虾备好料没吃，单等第二天中午加藤来店里吃午饭。加藤一来，李启贵就把做好的几种龙虾菜肴让日本女服务员端上来了。其中就有这道"空心龙虾球"。此菜的原料是大龙虾、猪肥膘肉、南荠、猪皮、火腿、青豆、面包粒、葱姜、鸡蛋、精盐、料酒、胡椒粉、水淀粉、食用油等。将龙虾宰杀好，头尾要完整，用水焯透，分置于鱼盘的前后部位。龙虾肉去筋皮，与肥膘肉、南荠一样均切成粒，放入盆中，加精盐、胡椒粉、鸡蛋液、葱姜油、水淀粉拌匀成虾胶。肉皮去净油和毛，切成条状，加入适量的清水和葱姜、火腿、胡椒粉，上屉将肉皮蒸到出胶性即可。然后加料调好味，加入青豆，冷却后切成2.5厘米的方丁。面包去皮切成小粒。将虾胶均匀地分成12份，每份包上适量皮胶，粘上面包粒，制成虾球，下入四五成热的油中炸成金黄色，捞出摆在虾头和虾尾之间即成。上桌的时候，用餐刀切开金黄的虾球，顿时香气四溢，汤汁横流，吃起来外焦里嫩，味美鲜香。加藤先生连连称赞。

有意思的是，第二天他再来的时候，李启贵并没有拿大龙虾给他做，而是改

By this day, Li Qigui worked in Osaka for less than a month. The friendship between Chinese and Japanese chefs also started at that moment.

Soon after that, Li Qigui showed another unique skill in front of Japanese chefs. The Chinese banquet in Taoyuan Restaurant was of relatively high grade, and Li Qigui had creatively introduced a new dish "Hollow Lobster Balls".

When it came to making seafood dishes, Japanese chefs were somewhat boastful. Japan is an island country in the sea and naturally has more seafood. Seafood sashimi is best dish in Japanese cuisine. They wondered whether Li Qigui, a Chinese chef from Taifeng Restaurant in Beijing, was good at making seafood. Without speaking out this doubt directly, Mr. Katô asked staff to buy a 3 kg lobster from the seafood market. He walked into the kitchen with the lobster of long shrimp antennas and said to Li Qigui with a smile, "Mr. Li, please have this for dinner." Li Qigui happily took it over and thought that the boss would invite him to have this for dinner? It's too expensive! But on second thought, something was wrong, and there must be something behind it! He wanted Japanese chefs to see if he could cook lobster and seafood.

Then, Li Qigui smiled at ease. Since preparing for the IKA, Master Wang Yijun had taught him how to make seafood. As a winner of the gold medal on the IKA, he took the "lobster test" before him simply as a piece of cake. That night, Li Qigui prepared the lobster and did not cook it. He waited for Katô to come to the restaurant for lunch at noon the next day. As soon as Katô arrived, Li Qigui asked the Japanese waitress to serve several lobster dishes prepared by himself. One of them was Hollow Lobster Balls. The main ingredients of this dish were lobster, fat pork, water chestnut, pig skin, ham, green beans, bread grains, scallion and ginger, eggs, refined salt, cooking wine, pepper, water starch, edible oil, etc. Kill the lobster, keep its head and tail complete, blanch it thoroughly with water, and place them on the front and rear parts of the fish plate. Remove the skin of lobster, cut it into particles like fat pork and water chestnut, put them into a basin, add refined salt, pepper, egg liquid, fried scallion-ginger oil and water starch, and mix well to form minced shrimp. Remove fat and hair from the pig skin, cut it into strips, add appropriate amount of clear water, scallion, ginger, ham and pepper, and steam the skin in a drawer until it had glue. Then add seasonings and green beans, cool and cut into 2.5 cm cubes. Peel the bread and cut it into small pieces. Divide minced shrimp evenly into 12 portions, wrap each portion with appropriate amount of skin glue, stick bread grains, make shrimp balls, fry in hot oil when its temperature reaches 40% to 50% of boiling point to be golden yellow, take out and place them between head and tail. When serving the table, cut the golden shrimp balls with a knife, and immediately the aroma overflowed and the soup flowed. It tasted delicious and fresh, tender with a crispy crust. Mr. Katô praised again and again.

Interestingly, when he came again the next day, Li Qigui did not cook the big lobster for him. Instead, he cut the smallest shrimps into paste and made it into hollow lobster balls for him to taste. Katô said it tasted great. Li Qigui asked him what shrimp meat was used. Katô said this was the lobster I gave you! Li Qigui told him with a smile, "This is made of very small shrimps by chopping into paste!" Katô was surprised when he heard this. He patted Li Qigui's wrist repeatedly and said, "Great, great, I am very satisfied, and in this way, I can make a lot of money!"

用最小的虾仁斩成蓉，再做成空心龙虾球请他品尝。加藤一吃说味道很棒。李启贵便问他这是什么虾肉做的？加藤说："这就是我给你的龙虾呀！"李启贵笑着告诉他："这是用很小的虾仁剁成蓉做出来的！"加藤闻言很吃惊，他连连拍着李启贵的手腕说："了不起，了不起，我很满意，这样我就能赚到很多钱了！"

在加藤看来，李启贵大师的美食创意好像无处不在。有加藤的支持，李启贵的厨艺天赋被进一步激发出来。南海假日酒店经常承接一些每桌餐标二三十万日元的高档席面，李启贵开出的菜谱都是比较难做的，一般日本厨师做不出来。恰恰因为这样，物以稀为贵，李大师的菜品备受客人的青睐。比如"芙蓉鲍片"这道菜，是用日本的大活鲍鱼再加上松茸为原料，配料是鸡里脊、鸡蛋清和青豆，调料包括上好清汤、水淀粉、精盐、葱姜油、料酒、胡椒粉和姜汁。先将鲜鲍鱼片成大片，鸡里脊去筋制成蓉，然后加入少量的水淀粉和少量精盐，在三成热的温油中吊成大片备用。天然黑松茸切薄片，汤勺上火，放入清汤，加入精盐、胡椒粉、吊好的鸡片、鲍鱼片、松茸片、姜汁、料酒，汤开后勾薄芡，放入少量葱姜油，装盘点上青豆即可。这道菜看上去鸡片雪白、鲍鱼片微黄滑嫩，野生松茸片质地脆嫩，黑白黄色泽分明，鲜香适口。

后来，李启贵在大阪的声名日隆，许多餐饮业的大厨都要来学李启贵的厨艺。加藤立刻把这个现象转化为商机，他说："要想听李启贵大师讲课学习，想吃他做的菜，都没关系，到我的店里来，每人25000日元的餐标。" 结果李启贵展示厨艺，加藤大赚了一笔。

当然，李启贵的"传道授业解惑"也是有限度的。日本餐饮业当时有人正在研究怎么把中国的龙须面弄上工厂的流水线，一直没有成功。这个厂长就通过加藤先生找到李启贵，说想请他去大阪南边一个岛上的面点工厂看看，同时把抻面的技术在那儿教教他们。李启贵是一名有觉悟的名厨，公派来日本执行组织交给的合约任务，不能做超越合约之外的事，否则就违规了。出于友好交往的目的，李启贵答应给他们现场展示一次，但是只可以拍照，不能录像。加藤问可否录一遍视频？可以付给他30万人民币。李启贵说："你给我300万人民币我也不要，也不能录。"几天后，李启贵在加藤的陪同下到了这家面点工厂，现场表演制作了豌豆黄、芸豆卷、龙须面，日方只是拍了些照片，中国顶级面点的手艺没有被偷学了去。

In Katô's view, Master Li Qigui's creativity in cooking seemed to be everywhere. With Katô's support, Li Qigui's cooking talent was further stimulated. Holiday Inn Nankai often undertook some high-grade dish packages with a price of 200,000 to 300,000 yen per table. The recipes prescribed by Li Qigui were relatively difficult to prepare, and ordinary Japanese chefs couldn't make it, which made Master Li's dishes more favored by customers. For example, the dish Lotus & Abalone Slices took Japanese live abalone plus tricholoma matsutake as main ingredients, chicken tenderloin, egg white and green beans as ingredients and excellent clear soup, water starch, refined salt, fried scallion-ginger oil, cooking wine, pepper and ginger juice as seasonings. First, cut the fresh abalone into large pieces, remove the tendons of chicken tenderloin and mince into paste, then add a small amount of water starch and a small amount of refined salt, and fry the fresh abalone slices into large pieces in 30% hot warm oil for later use. Cut natural black tricholoma matsutake into thin slices, add clear soup in the spoon and heat, add refined salt, pepper, fried chicken slices, abalone slices and tricholoma matsutake slices, ginger juice and cooking wine, thicken the soup after boiling, add a small amount of fried scallion-ginger oil, and add green beans after placing on the plate. With white chicken slices, slightly yellow and tender abalone slices, and crisp and tender wild tricholoma matsutake slices, this dish looked clear, and tasted fresh and delicious.

Later, Li Qigui was growingly well-known in Osaka, and many chefs in the catering industry came to learn Li Qigui's cooking skills. Katô immediately turned this phenomenon into a business opportunity. He said, "If you want to listen to Master Li Qigui's lectures and study from him or want to eat the dishes he cooks, please come to my restaurant and pay 25,000 yen/person/meal." As a result, Li Qigui showed his cooking skills and Katô made a lot of money.

Of course, Li Qigui didn't give lectures without limit. At that time, some people in Japanese catering industry were studying how to introduce Chinese Longxu Noodles on the assembly line of a factory, but they did not succeed. The factory director found Li Qigui through Mr. Katô and invited him to visit a noodle factory on an island south of Osaka, inviting him teach them noodle pulling techniques there. Li Qigui realized that he was a famous chef with consciousness. He was sent to Japan to carry out the contract tasks assigned by his company. He could not do anything beyond the contract; otherwise he would violate the law. For the purpose of friendly exchanges, Li Qigui promised to show them on the spot, but they could only take photos instead of recording videos. Katô asked if he could record a video. Li Qigui said, "No, recording would not be allowed even if you give me RMB 3 million." A few days later, accompanied by Katô, Li Qigui went to the noodle factory and performed on the spot to make pea cake, kidney bean rolls and Longxu Noodles. Japan only took some photos. China's top noodle skills were not stolen.

Intentionally, Li Qigui always held the attitude of teaching benefits teachers as well as students, learning from Japan's advanced catering management mode modestly, exploring the future development trend of the catering industry, and making corresponding knowledge and ideological reserves for the development of Chinese catering service industry while he worked in Japan. This kind of foresight was not possessed by every employee in the catering industry.

In this way, Li Qigui had worked in Osaka for more than two years, not only making money for

李启贵的有心之处还在于，他在日本工作，始终抱着教学相长的态度，虚心学习日本先进的餐饮管理模式，探究餐饮业今后的发展动向，为中国餐饮服务业的发展做好相应的知识储备和思想储备。这种超前意识，并非每一位餐饮业从业人员都具备。

就这样，李启贵在大阪工作了两年多，既为日方老板赚了钱，也为国家创造了外汇，还为酒店培养了一批中餐厨师，使他们掌握了一定的中餐烹饪技能，为中日饮食文化交流做出了贡献。1992年合约结束，他要回国了，老板很不愿让他走，连问了他三个问题："第一，您为什么非要回国？是因为钱少吗？钱少我可以给您加钱。是因为居住的条件不好吗？我可以给您换大房子。还是说在日本一个人过得孤单寂寞？我可以给您找一个日本老婆。"李启贵告诉他，这些原因都不是。他回国是因为合约期满，他要回单位工作，另外他的老父亲病重，必须回去陪老人家看病。日方老板只好恋恋不舍地和李启贵道别，感谢他的精湛厨艺和真诚付出，同时也对他高贵的人品表示由衷的敬意。李启贵回国后两个多月，他的老父亲李春秀就因心衰去世了，当时，李启贵就陪在父亲的病床前。

转过年来，1993年9月，李启贵在师父王义均大师的口传心授之下，又挽挽袖面参加了第三届全国烹饪大赛。他制作的"空心龙虾球""茉莉凤蓉竹荪汤"，造型精美、色泽雅致、醇香鲜厚，得到现场评委一致好评，夺得金牌，被授予"全国优秀厨师"称号。此后不久，又一光荣使命落在了李启贵的肩上，他高票当选为迎奥运北京烹饪大赛评委。

1994年，李启贵参加了由中国烹饪协会举办的"二国五方"厨艺大赛，再次人前显贵鳌里夺尊，以冷拼"二龙戏珠"和刀工热菜荣获金奖第一名。随后，他被北京商学院聘为客座教授，被北京市劳动局聘为烹饪专业高级讲师。

1997年7月1日，香港回归，举国欢庆。同年11月，应香港京华国际大酒店董事长荣世浩的邀请，李启贵来香港交流饮食文化，增进彼此了解。此时的李启贵已经是泰丰楼饭店的副总经理兼总厨师长，被聘为香港回归烹饪大赛的评委。所以来香港之前，李启贵心里就一直在盘算，此次赴港交流，到底要给见多识广的香港同胞展示些什么耳目一新的东西？

思前想后，李启贵决定还是以看家本领"京派鲁菜"示人，推出以"鲁

the Japanese boss, but also creating foreign exchange for the country. He had also trained a group of Chinese chefs for the hotel, enabling them to master certain Chinese cooking skills and contributing to the exchange of Chinese and Japanese food culture. When the contract ended in 1992, he was going back to China. The boss was very reluctant to let him go. He even asked him three questions, "First, why do you have to go back to China? Is it because of money? I can give you more money if you want. Is it because of poor living conditions? I can provide a big house for you. Or can I find you a Japanese wife if you live alone in Japan?" Li Qigui told him that none of these reasons were true. He returned to China because the contract expired and he had to return to work in his unit. In addition, his old father was seriously ill and had to go back to take his father to see a doctor. The Japanese boss had to reluctantly say goodbye to Li Qigui, thanking him for his exquisite cooking skills and sincere efforts, and also expressing heartfelt respect for his noble character. More than two months after Li Qigui returned home, his old father Li Chunxiu died of heart failure. At that time, Li Qigui was beside his father.

In September 1993, Li Qigui took part in the 3rd IKA under the encouragement and instruction of Master Wang Yijun. The "Hollow Lobster Balls" and "Dictyophora Soup with Jasmine & Celosia Cristata" made by Li Qigui were exquisite in shape, elegant in color, and mellow, fresh and thick in taste, winning unanimous praise from the judges on the spot. He won the gold medal and was awarded the title of "Chinese National Excellent Chef". Shortly thereafter, another glorious mission fell on Li Qigui. He was elected with a majority of votes as a judge of the Beijing Cooking Competition for the Olympics.

In 1994, Li Qigui took part in the two-country and five-party cooking competition organized by the China Cuisine Association. Once again, with outstanding performance and unique cooking skills, he won the first gold medal via the assorted cold dish "Two Dragons Play with A Pearl" and a hot dish showing special cutting skills. Later, he was employed as a visiting professor by Beijing Commercial College and a senior lecturer in cooking by Beijing Labor Bureau.

On July 1, 1997, Hong Kong returned to China, which the whole country celebrated. In November of the same year, at the invitation of Rong Shihao, Chairman of the Metropark Hotel Kowloon in Hong Kong, Li Qigui came to Hong Kong to exchange food culture and enhance mutual understanding. Then, Li Qigui was already the Deputy General Manager and Chief Chef of Taifeng Restaurant and was employed as the judge of the Hong Kong Return Cooking Contest. Therefore, before coming to Hong Kong, Li Qigui had been thinking about showing what refreshing things to the well-informed Hong Kong compatriots during this exchange.

After careful deliberation, Li Qigui decided to show them his unique skills of Beijing-style Shandong cuisine and launch special dishes with the theme of Meeting Metropark with Shandong Cuisine, with more than 60 kinds of dishes. As soon as the news came out, 23 Hong Kong medias, including Hong Kong Ta Kung Pao, Hong Kong Commercial Daily, Sing Tao Daily, Ming Pao, Apple Daily, East Weekly, Oriental Daily News, Hong Kong Wen Wei Po, The Standard, Hong Kong ATV, HKTVB and others, reported one after another, causing a sensation.

All kinds of classic dishes and special new products were presented under the reputation of

馔会京华"为主题的特色美食汇,菜品多达六十余种。消息一出,包括香港大公报、香港商报、星岛日报、明报、苹果日报、东周刊、东方日报、香港文汇报、虎报、香港亚视、香港无线等23家香港媒体纷纷报道,可谓轰动一时。

在"鲁馔会京华"的盛名之下,各种经典菜肴、特色新品可谓包罗万象,异彩纷呈。其中的名菜有红扒通天鱼翅、红扒通天海虎翅、烩乌鱼蛋、扒燕菜卷、芙蓉鲍片、红扒驼掌、兰花驼掌、葱烧海参、龙须全蝎、三丝鱼翅、辽东海参汤、山东海参、鸡蓉竹荪、鲍丝什锦羹、清汤燕窝、白扒猴头蘑、北菇菜心、烧四宝、芫爆鲜鱿、海参扒鸡腿、油焖大虾、干燸大虾、炸烹大虾、垮炖目鱼、炸板鱼、糟熘鱼片、醋椒鱼、糟熘三白、油爆双脆、扒牛肉条干、干燸活鱼、炸两样、龙须海王鲍、海参扒鱼肚、两吃大虾……冷菜包括玻璃蝴蝶虾、麻仁杏干肉、五香熏鱼、珊瑚卷心菜、虾子冬笋、蜜汁红枣、油焖香菇、麻仁干钱肉。

点心推出雪花龙须面、罘山拉面、三鲜四喜蒸饺、萝卜丝饼、麦穗包、临清烧卖、门钉馒头、肉酥饼、炸烹虾带皮等。此外还有一些海鲜菜肴。当时接待的贵宾基本都是香港烹饪协会永远会长杨维香、东西促进会会长甄文达等香港社会知名人士,以及各行各业的老板、董事长。李启贵带来的菜品受到了各界的好评。

李启贵此行更重要的任务是做好"鲁馔会京华"主题餐饮的推广培训工作。怎么做呢?港方要求李启贵大师把京华国际酒店的厨师、服务员、采购部,包括管理人员,通通培训。这给李启贵出了一道难题,因为工作岗位不一样,人员分工不同,可又是完成同一个项目,得做通盘考虑,怎么培训才能既高效又明晰呢?李启贵想,这就如同古时候排兵布阵一样,弓弩手负责压住阵脚,探子手负责通风报信,大将负责打头阵,粮草官负责保障军需。各司其职,令行禁止,形成"一棵菜""一盘棋"。

拿定主意,李启贵开始安排起他的大阵。他让工作人员摆了一个丁字台。李启贵居中稳坐,左边是服务员,右边是厨师。身后两侧立起两个大货架子,一边是原料,一边是炊具。比如做一个红扒海虎翅,首先李启贵要跟左边的厨师交代清楚,要备水发海虎翅多少;然后对右边的服务员说这个菜的卖价是多少、利润和价格是多少、用哪块盘子装菜。李启贵举起盘子说,这个是用12寸的扒盘。一道道菜挨个儿这么培训,但是李启贵讲得清清楚楚,一百多人听得

Meeting Metropark with Shandong Cuisine. Among them, the famous dishes included Braised Supreme Shark's Fin, Braised Supreme Tiger Shark's Fin, Stewed Black Fish Eggs, Braised Bird's Nest & Vegetable Rolls, Lotus & Abalone Slices, Braised Camel's Palm, Braised Sea Cucumbers with Orchid, Camel's Paw & Scallion, Whole Scorpion with Gracilaria, Shark's Fin with Three Shredded Ingredients, Liaodong Sea Cucumber Soup, Shandong Sea Cucumbers, Bamboo Fungus with Minced Chicken, Assorted Soup with Shredded Abalone, Clear Soup with Bird's Nest, Stewed Hericium Erinaceus, Chinese Cabbage with Northern Guangdong Mushroom, Braised Four Ingredients, Quick Fried Squid with Coriander, Braised Chicken Legs with Sea Cucumbers, Braised Prawns, Simmered Prawns, Fried Prawns, Stewed Cuttle Fish, Fried Flatfish, Sauteed Fish Fillets with Wine Sauce, Fried Fish with Vinegar and Pepper, Sauteed Chicken, Fish and Bamboo Shoots with Rice Wine Sauce, Stir-Fried Double Crisp, Stewed Beef Strips, Simmered Live Fish, Fried Two Ingredients, Sauteed Abalone with Gracilaria, Braised Fish Maw with Sea Cucumber, Prawns Cooked in Two Styles ... Cold dishes included Glass Butterfly-like Shrimp, Dried Apricot Meat with Sesame Kernel, Spiced Smoked Fish, Coral-like Cabbage, Shrimp and Winter Bamboo Shoots, Honeyed Red Dates, Braised Mushrooms with Oil, and Dried Meat with Sesame Kernel.

Snacks included Snowflake Longxu Noodles, Fushan Noodles, Sixi Steamed Dumplings Stuffed with Shredded Seafood, Cake with Shredded Radish, Wheat Head-like Buns, Linqing dumplings, Door Nail-like Steamed Bread, Crispy Meat Cake, Fried Shrimp with Skin, etc. In addition, there were some seafood dishes. At that time, the distinguished guests received were basically social celebrities in Hong Kong such as Yang Weixiang, Permanent President of Hong Kong Cooking Association, and Zhen Wenda, President of the East-West Promotion Association, and bosses and chairmen from all walks of life. The dishes cooked by Li Qigui had received favorable comments from all walks of life.

The most important task of Li Qigui for that trip was to do a good job in the promotion and training under the catering theme of Meeting Metropark with Shandong Cuisine. How to do a good job? Master Li Qigui was asked to train all the chefs, waiters and purchasing departments of Metropark Hotel Kowloon, including management personnel, which posed a difficult problem for Li Qigui. Due to different jobs and different division of labor among personnel, it is necessary to make overall consideration to complete the same project. How could the training be efficient and clear? Li Qigui thought it just like the layout of troops in ancient times. Different positions should be responsible for different duties and every order should be executed without fail. A systematic situation should be developed properly.

Once making up his mind, Li Qigui began to arrange his layout. He asked the staff to set up a T-table. Li Qigui sat firmly in the middle, with the waiter on the left and the cook on the right. Two large shelves were erected on both sides behind them, with main ingredients on one side and cooking utensils on the other. For example, to cook Braised Supreme Tiger Shark's Fin, Li Qigui firstly explained to the chefs on the left how many sea tiger shark's fins should be prepared with water, then told the waiter on the right the selling price of this dish. What was the profit and price, and which plate to use for the dishes? With a 12-inch plate in his hand, Li Qigui said that this dish should be placed on the plate. One dish by one dish, Li Qigui spoke clearly to ensure more than 100 people

明明白白。别人一天都做不完的重要培训，李启贵用了两多钟头就完成了。

当然，光说不练假把式。为了推广京派鲁菜，李启贵在香港的亚洲电视台打头阵，表演了四道名菜。第一道是"芙蓉鲍片"。这道菜吊片是关键步骤，难度很大，蛋清多了鲍片就起泡儿，蛋清少了团粉多了，鲍片就发硬。所以对用料和火候的把握要求都很高。第二道是"红扒通天海虎翅"。"扒菜"是个大翻锅的技术，要把锅子晃动起来，然后把菜抛起来空中接住，最后落到锅里，装在盘里，表演起来惊险刺激。第三道菜是龙须全蝎。李启贵不用罐头的，用的是碧绿的鲜龙须。把蝎子炸好放在鲜龙须菜上。炸蝎子是京鲁菜的名菜，按传统中医的说法，蝎子有息风止疼、通络净血的作用。炸蝎子的关键在于活蝎子的处理上。活蝎子拿来，搁在一个缸里或者一个小口的罐子里，放上凉水，放上盐，把盐搅化了，把这活蝎子给爆腌了。等做菜的时候再捞出来，下锅炸制。李启贵大师炸蝎子的手法跟别人也不一样，锅里的油到了六七成热，就开始下锅炸蝎子，因为油温太热就炸煳了。把蝎子下到锅里一炸，旋即捞出。第二次油温升高，再下去炸，得把蝎子的肚子炸爆了，才能达到酥香而不苦煳的效果。吃蝎子关键是吃钩子，这也是蝎子的剧毒所在。炸好的蝎子一是脆，二是醇香，三是吃炸全蝎，它的爪子的酥度、身子的酥度和钩子的酥度是不一样的，齿颊能同时品尝到三种不同的感觉，非常奇特。吃完一只蝎子，再吃一根荷兰进口的龙须菜，清新爽口。第四道是雪花龙须面，现场和面现场抻，13扣一气呵成。只见李启贵右手向斜上方空中一挑，面长两米有余，万线银丝，根根透风，条条劲健。连见多识广的电视台主持人都惊喜非常，第一个鼓掌叫起好儿来。

"千金容易得，一将最难求"。像李启贵这样在餐饮业能够"六场通透，文武昆乱不挡"的人才，更是凤毛麟角。北京王府井附近的一家豪华酒店，早早就看到了李启贵大师的价值。先是聘他当烹饪顾问，在李启贵休息的时候，从泰丰楼用车把他接去帮助指导业务，设计餐饮结构。李启贵从香港载誉归来，他们便来了个"竹筒倒豆子"，明确表示要请李启贵去做总厨师长。没想到，泰丰楼的上级单位说什么也不放，这可让李启贵左右为难：去，自然好，那是个更广阔的舞台，自己年富力强，事业正在巅峰状态，可以大显身手；可是上级单位待自己不薄，把他当成擎天白玉柱、架海紫金梁，树成标杆，让全区饮食行业的从业人员学习。榜样，大家学习的榜样，能走吗？

understand it thoroughly. Li Qigui completed the important training in more than two hours, which others could not finish in one day.

Of course, action speaks louder than words. In order to promote Beijing-style Shandong cuisine, Li Qigui took the lead in Hong Kong's ATV station to perform four famous dishes. The first was Lotus & Abalone Slices. The key step of this was slicing, which was very difficult. If there was more egg white, the abalone slices would bubble. If there was less egg white with more powder, the abalone slices would harden. Therefore, there were very higher requirements for materials and temperature. The second was Braised Supreme Tiger Shark's Fins. Braising was a technique of turning the pot upside down, i.e., shaking the pot, then throwing up the dish and catching it in the air. Finally, it should fall into the pot and be placed on the plate. The performance was breathtaking and exciting. The third was Whole Scorpions with Gracilaria. Li Qigui did not use green fresh gracilaria instead of canned one. Fry the scorpions and put it on the fresh gracilaria. Fried Scorpions was a famous dish in Beijing-style Shandong cuisine. According to traditional Chinese medicine, scorpions had the functions of relieving pain, dredging collaterals and purifying blood. The key to frying scorpions lies in the handling of live scorpions. Take live scorpions, put them in a jar or a small jar, pour cold water, add salt and solve it so as to marinate live scorpions. When cooking, take it out and fry it in a pan. Master Li Qigui fried scorpions in a different way, that's, starting to fry scorpions in the pan when the oil in the pan was heated to 60% to 70% of its boiling point, since scorpions would be crisp if the oil temperature was too higher. Fry scorpions in the pan and take it out immediately and then fry them again when the oil temperature rose until that the scorpion's belly was blown up so as to achieve the effect of crisp fragrance without bitter and crisp taste. The key to eating scorpions was to eat hooks, which were highly toxic. The fried scorpions should be crisp, mellow and delicious. The crisp degree of its claws, body and hooks were different. You could have three different feelings at the same time, which was very special. After eating a scorpion, have a piece of fresh gracilaria imported from Holland, which was fresh and refreshing. The fourth dish was Snowflake-like Longxu Noodles, which was pulled on the spot with kneading the dough also on the spot and 13 folds were pulled at one go. It's seen that Li Qigui lifted his right hand in the air in the oblique direction, the noodles stretched more than two meters, like ten thousand threads of silver, ventilated and sound between each other. Even the well-informed TV host was very surprised to applaud and shout as the first one.

It's easy to make money but hard to get a talent. It's rare in the catering industry to have such talents as Li Qigui who were outstanding in every aspect. A luxury hotel near Wangfujing in Beijing valued Master Li Qigui for a long period and hired him as a cooking consultant at first. When Li Qigui had a break, they picked him up from Taifeng Restaurant to help guide the business and design the catering structure. When Li Qigui returned from Hong Kong with a good reputation, they made it clear that they would invite Li Qigui to be the chief chef. Unexpectedly, the high management of Taifeng Restaurant did not let him go regardless of any reasons, which put Li Qigui in a dilemma: Of course, it was a broader stage to display his talents in his prime at the peak of his career if he went there; meanwhile, the high management had treated him well and regarded him as a backbone and a model for the employees in the catering industry in the whole region to learn from. How could he leave as a model?

第五回

Chapter Five

身教为先 盛宴亲烹三道菜
天伦布阵 六大名厨闹京都

———

1998年初，李启贵完成香港的表演交流任务，率队回到了泰丰楼。此时，他既是泰丰楼的副总经理兼总厨师长，又是王府井天伦王朝酒店的顾问。

提起1990年开业的天伦王朝酒店，北京人几乎无人不知无人不晓。1997年，天伦王朝酒店的上级单位是北京国际信托投资公司。作为北京的对外窗口单位，天伦王朝酒店的各项经营标准必须与国际接轨，这就需要有行业内的顶尖级人物来鼎力加盟，专业的事要由专业的人来做。

李启贵在餐饮业连续三届是北京市的"拔尖人才"。作为个中翘楚，他自然受到了业内同行的高度关注。先下手为强，天伦王朝酒店旋即把美食顾问的聘书送到了李启贵的手上。受人之托，忠人之事，于是李启贵就利用休息时间，到天伦王朝酒店指导餐饮服务，双方相处得非常合拍，非常愉快。酒店领导就有了"得陇望蜀"的念头，公司开会一商量，干脆挖人！咱们把李启贵大师请到天伦王朝集团做总厨师长！此时的李启贵左右为难：去，自然好，那是个更广阔的舞台，自己年富力强，事业正在巅峰状态，可以大显身手；可是上级单位待自己不薄，说走就走，于心不忍。果不其然，宣武区饮食服务公司的领导跟李启贵说："启贵，你看看，这红头文件上盖了7个大红章，是吧？这是号召公司全体从业人员向李启贵同志学习的红头文件。我们把您当成擎天白玉柱、架海紫金梁，树成大标杆，让全区饮食行业的从业人员学习。你要是走了，大伙儿学谁啊？"李启贵也觉得赵书记的话说得在理。没想到，天伦王朝那边志在必得，锲而不舍的劲儿还挺大，明确表示，李大师来天伦王朝工作，每天负责车接车送，酒店要是给职工分房，第一套就是您的。

又过了几天，宣武区的区委副书记兼组织部长郑文奇知道了这件事。他的看法却与众不同。他认为不放人是不对的，"在任何一个工作岗位上，都是为人民服务，有一个更重要的岗位需要你，他们应该同意。这个工作呢，我来负责做。"郑书记的话，让李启贵心里热乎乎的。郑书记言而有信，没过几天公司就来信儿了，同意放人。于是1998年下半年，李启贵正式调入天伦王朝酒店任行政总厨。

By Setting Examples, He Cooked First Three Dishes for All Important Banquets
With Exquisite Layout and Design, Six Famous Chefs Competing in Beijing

At the beginning of 1998, Li Qigui completed the performance exchange mission in Hong Kong and led the team back to Taifeng Restaurant. At that time, he was not only Deputy General Manager & Chief Chef of Taifeng Restaurant, but also the Consultant of Sunworld Dynasty Hotel in Wangfujing, Beijing.

When it comes to the Sunworld Dynasty Hotel, which was opened in 1990, almost everyone in Beijing knows it. In 1997, the high management of Sunworld Dynasty Hotel was Beijing International Trust and Investment Company. As Beijing's external window unit, the operating standards of Sunworld Dynasty Hotel must be in line with international standards, which requires top figures in the industry to join in and to do professional things by professionals.

Li Qigui had been a "top-notch talent" in Beijing for three consecutive sessions in the catering industry. As an outstanding figure, he naturally received high attention from his peers in the industry. To take the initiative was to gain the upper hand. Sunworld Dynasty Hotel immediately sent Li Qigui the letter of appointment of food consultant. Being entrusted and loyal to Sunworld Dynasty Hotel, Li Qigui took advantage of his rest time to guide catering services at Sunworld Dynasty Hotel. The two sides got along very well and enjoyed themselves very much. The hotel leaders had insatiable desires. As soon as the company held a meeting to discuss it, they simply decided to hire him and invite Master Li Qigui as the chief chef of Sunworld Dynasty Hotel! At that time, Li Qigui was in a dilemma. Naturally, it's a broader stage to display his talents if he went there since he was in his prime and at the peak of his career. However, the high management treated him well and he was reluctant to leave former restaurant. As expected, the leader of Xuanwu District Catering Service Company said to Li Qigui, "Qigui, look at this official document with seven red seals, right? This is an official document calling on all employees of the company to learn from Comrade Li Qigui. We regard you as a backbone and model for the employees in the catering industry in the whole region to learn from. If you leave, who would everyone follow?" Li Qigui also felt that the Secretary's words were reasonable. Unexpectedly, the Sunworld Dynasty Hotel was determined to get him. It was made clear that Master Li would be picked up every day and even enjoyed the first position in case of house allocation if he came to work in Sunworld Dynasty Hotel.

A few days later, Zheng Wenqi, Deputy Secretary and Organization Minister of Xuanwu District, learned about this. His view was different. He thought it was wrong if not letting him go. "It's all to serve the people regardless of positions. They should agree with you if there is a more important job available. I will be responsible for negotiation." Secretary Zheng's words warmed Li Qigui. Secretary Zheng kept his word, and within a few days the company wrote to Li Qigui and agreed with him about his leaving. So in the second half of 1998, Li Qigui officially transferred to Sunworld Dynasty Hotel as the executive chef.

此时的天伦王朝还是家四星级酒店，酒店的餐饮主打是"四季厅"，经营所谓"四方风味"的菜肴，餐饮格局既不清晰也不合理。李启贵决定从根儿上解决这个问题。路子走对了，所有问题便会迎刃而解。

他山之石，可以攻玉。北京其他酒店的成功经验是否可以借鉴呢？李启贵上任伊始，就请上酒店分管餐饮的副总经理安学峰和餐饮总监毕荣江一起到北京友谊宾馆取经考察。

当时的友谊宾馆有4个大厨房，一个粤菜的，一个川菜的，一个淮扬菜的，还有个多功能自助餐厅。餐饮结构基本合理。看完之后，李启贵心里基本有了路数。他就跟那两位领导商量："您看，咱们的菜叫四方菜，餐厅叫四季厅，听起来都太不专业，因为特色不突出。比如川菜，比如粤菜，一定要特色鲜明，才能脍炙人口。"天伦王朝的安总和餐饮总监也觉得，人家友谊宾馆的川菜格局很成功，守着炒锅的就是个四川人。

要找正宗川味，必须先去成都。揣着这个思路，走马上任的李启贵就和副总经理安学峰俩人够奔四川成都了。耳听为虚，眼见为实，亲自品尝才算做到心里有底。为了找到合适的川菜厨师，他们在成都先后品尝了很多家川菜馆，其中既有朋友事先介绍的，也有憨着机缘巧合偶遇的。临出发的时候，在天伦王朝负责装修的谢宇是个四川人，他说认识一位成都的川菜老师傅，手艺不错，热情地介绍他们去试试。按图索骥，他们找到了谢宇说的这位川菜师傅，年纪在四十五六岁。李启贵点了几道川菜的代表菜，比如宫保鸡丁、干煸牛肉丝、鱼香肉丝、麻婆豆腐等，就是要实际检验一下厨艺高低。因为李启贵曾经在北京西绒线胡同的四川饭店学过半年多的川菜，所以其中的门道儿他熟稔于心。上来的第一道菜是宫保鸡丁，结果发现这个菜不是过油炒的，而是卧油炒的，油大了。其次，肉菜改刀，讲究"丝儿细、片儿薄、丁儿匀"，这份宫保鸡丁显然鸡丁切得太碎小了，而且并不匀实，可见刀工一般。俩人一交换意见，安总也这么看，觉得一个厨师的操作习惯很难改变，为了保证天伦王朝饭店高水平的餐饮质量，不得不另做打算。

又转了其他几个饭馆，也是这样。或许成都的川菜和北京的川菜还真是不大一样，俩人乘兴而来，无功而返。在回来的火车上，安学峰跟李启贵说："李大师你有什么想法？"李启贵说："我的想法是这样，川菜是在辣的基础上有花椒有麻的感觉，因为川贵一带比较潮湿，花椒呢有祛湿的功效。咱们北方人要是接受辣，我认为还是湘菜比较适合。湘菜是香辣，而且不甜不麻。"

At that time, the Sunworld Dynasty Hotel was still a four-star hotel. Its catering service was mainly served at the Four Seasons Hall, which sold the so-called All Flavor dishes. The catering pattern was neither clear nor reasonable. Li Qigui decided to solve the problem fundamentally. All problems would be properly solved if they adopted a right way.

Stones form other hills may serve to polish the jade of this one. Could the successful experiences of other hotels in Beijing be used for reference? At the beginning of Li Qigui's term of office, he invited An Xuefeng, Deputy General Manager in charge of catering of the Hotel, and Bi Rongjiang, Catering Director, to visit Beijing Friendship Hotel and learn from the experience.

At that time, Beijing Friendship Hotel had four large kitchens, one for Cantonese cuisine, one for Sichuan cuisine, one for Huaiyang cuisine and a multi-functional cafeteria, which was basically reasonable in structure. With this knowledge, Li Qigui basically had an idea. He discussed with the two leaders, "As you see, our dishes are called All Flavor ones and the restaurant is called Four Seasons Hall. It sounds too unprofessional because the specialties are not outstanding. For example, in terms of Sichuan cuisine and Cantonese cuisine, we must have distinctive features in order to be popular." Directors An and Bi also felt that the Sichuan cuisine pattern of Beijing Friendship Hotel was very successful, and the chef of the frying pan was from Sichuan.

To find authentic Sichuan flavor, you should first go to Chengdu. With this train of thought in mind, Li Qigui, who just took office, and An Xuefeng, Deputy General Manager, were about to visit Chengdu, Sichuan. What you heard is not reliable but what you saw with your own eyes is. They tasted in person so as to rest assured. In order to find a suitable Sichuan cuisine chef, they tasted many Sichuan cuisine restaurants in Chengdu, some of which were introduced in advance by friends and some of which met by chance. At the time of departure, Xie Yu, who was in charge of decoration in the Sunworld Dynasty Hotel, was from Sichuan. He said that he knew an old Sichuan cuisine master from Chengdu with good cooking skills. He enthusiastically introduced them to have a try. Following up the cue, they found the Sichuan cuisine chef mentioned by Xie Yu, who was 45 or 46 years old. Li Qigui ordered several representative dishes of Sichuan cuisine, such as Kung Pao Chicken, Sautéed Shredded Beef with Chinese Pepper, Fish-Flavored Pork Shreds and Mapo Tofu in order to actually test his cooking skills. Because Li Qigui had studied Sichuan cuisine for more than half a year in the Sichuan Hotel in Xirongxian Lane, Beijing, he was familiar with the tips. The first dish served was Kung Pao Chicken. It was found that the dish was not fried with the main ingredients in the oil first, but fried with all ingredients together. So the dish was too greasy. Secondly, the meat and vegetables should be cut with emphasis on "thin shreds, thin slices and even dices". Obviously, the Kung Pao Chicken was made of too small and uneven chick dices, which showed the ordinary cutting skill. As soon as they exchanged views, Director An always felt as the same way. It was difficult to change a chef's operating habits. In order to ensure the high-level catering quality of the Sunworld Dynasty Hotel, he had to make other plans.

They visited several other restaurants, still no surprise found. Perhaps Sichuan cuisine in Chengdu and in Beijing were really different. They came on the spur of the moment and failed in return. On the train back, An Xuefeng said to Li Qigui, "Master Li, what idea do you have?" Li Qigui said, "I think that Sichuan cuisine has the taste of pepper as well as prickly ash, because Sichuan-Guizhou area is relatively humid and prickly ash has the effect of eliminating dampness. If we northerners accept spicy taste, I think Hunan cuisine is more suitable. Hunan cuisine is spicy without sweet nor numb taste."

有道是好事常多磨，好饭不怕晚。有时候思路一变，就会出现"柳暗花明又一村"的效果。在李启贵的身上，无论是精进厨艺，还是设计餐饮布局结构，勇于探索、敢于尝试、心灵手巧、沉着冷静的性格特点，给了他比别人更多的可能性和成功率。李启贵是首批"中国烹饪大师"，全国各省市、各菜系都有他的好朋友。所以他们二次南下，就奔了湖南长沙，去找湘菜泰斗聂厚忠大师。聂厚忠大师当时在长沙的蓉园宾馆工作，是湘菜响当当的领军人物。聂大师把他的得意门生介绍给李启贵，成为天伦王朝湘菜的执牛耳者。

天伦王朝集团旗下的38家五星级酒店都有粤菜厅，供应港式精品粤菜。北京天伦王朝酒店的粤菜厅叫"盛华瑄"。在盛华瑄主厨的大师名叫梁诚威，他是世界御厨、鲍鱼大王杨贯一的大徒弟，人称"威哥"。威哥与李启贵称兄道弟，关系很好。在梁诚威大师的鼎力相助之下，天伦王朝精品粤菜的声誉很好，生意也很红火，尤其是"威哥鲍鱼"，在京城可谓独领风骚。1999年9月29日，以盛华瑄餐厅为主，北京市委市政府在天伦王朝酒店举行颁授仪式，授予霍英东、李嘉诚、曾宪梓、惠京仔、李兆基、郑裕彤、郭炳湘、陈经纶、郭鹤年等社会名流"北京市荣誉市民"称号，随后的宴会隆重热烈，菜品鲜香美味，宾主尽欢。不久，香港著名爱国人士曾宪梓先生又在此举办了500人的重要宴会，也是以盛华瑄为主理，曾先生非常满意。

在港式精品粤菜取得成功的同时，李启贵又着手添置江南菜肴。江南菜肴属于中档菜，主要分布在长江流域以南的江浙一带，包括淮扬菜、苏州菜、无锡菜、上海菜、杭州菜等。上海菜偏甜，讲究浓油赤酱。杭州菜是鲜鱼水菜，口味以咸鲜为主。李启贵通过他的好朋友昆仑饭店总厨赵仁良介绍，请来了一位上海菜的厨师长，除了经营正宗的苏州、无锡名菜外，主要经营经过改良、更加适应北方人口味的上海菜。比如"河鲜汤"，就是用鲫鱼、河虾与河蟹作为主要原料，通过洗净、过油炸，然后煲汤，河鲜的味道极浓。

就这样，李启贵到了天伦王朝饭店之后，在请示主管领导和餐饮总监同意的情况之下，用了大约一年的时间，合理地调整了天伦王朝的餐饮结构，把地下一层原来自助餐厅的空间一分为二，一边增加了以"灵芝烤鸭"为特色的京派鲁菜；一边增加了以香辣为特色的湘菜厅。二楼原来的四季厅也来了个平分秋色，西边更名为"盛华瑄"，经营港式精品粤菜；东边成为上海菜肴的美食天地。四大菜系的格局就此形成了，也奠定了天伦王朝十几年餐饮结构的发展道路，使天伦王朝饭店在餐饮经营方面达到了鼎盛时期。有付出就有回报，

There is a saying that the realization of good things is usually preceded by rough goings and good meals are not afraid of being late. Sometimes when the train of thought changed, there was "a way out". As far as Li Qigui was concerned, whether he was diligent in improving cooking or designing the catering layout structure, he showed such characteristics as the courageousness of exploring and trying, ingeniousness and self-possession, which rendered him more possibilities and success rates than others. Among the first batch of "Chinese culinary masters", Li Qigui had good friends in a variety of cuisines across provinces and cities in China. So they went south for the second time and ran to Changsha, Hunan, to find Master Nie Houzhong, a leader in Hunan cuisine. Master Nie Houzhong was working in Rongyuan Hotel in Changsha at that time as a famous leader in Hunan cuisine. Master Nie introduced his favorite apprentice to Li Qigui, who became the leader of Hunan cuisine in the Sunworld Dynasty Hotel.

The 38 five-star hotels owned by Sunworld Dynasty Group all had Cantonese cuisine hall, which served excellent Hong Kong-style Cantonese cuisine. The Cantonese cuisine hall of Sunworld Dynasty Hotel Beijing was called "Shenghuaxuan Restaurant". The chef of Shenghuaxuan Restaurant was named Liang Chengwei, the first apprentice of world chef Master Yang Guanyi, who was good at cooking abalone, being called "Brother Wei". Brother Wei and Li Qigui were close brothers with a good relationship. Under the full help of Master Liang Chengwei, the excellent Cantonese cuisine of Sunworld Dynasty Hotel won a good reputation and its business was booming, especially Brother Wei Abalone, which had no rivals in Beijing. On September 29, 1999, taking Shenghuaxuan Restaurant as the major site, the Beijing Municipal Party Committee and Municipal Government held an award ceremony at Sunworld Dynasty Hotel, awarding celebrities such as Henry Fok, Li Ka-shing, Ceng Xianzi, Hui Jingzai, Li Zhaoji, Zheng Yutong, Guo Bingxiang, Chen Jinglun and Guo Henian the title of Honorary Citizen of Beijing respectively. The following banquet was grand and warm, with delicious dishes and pleased guests. Soon, Mr. Ceng Xianzi, a famous Hong Kong patriot, held another important banquet for 500 people here, also with Shenghuaxuan Restaurant as the major site, with which Mr. Ceng was very satisfied.

While having achieved success in Hong Kong-style fine Cantonese cuisine, Li Qigui had also started to serve Jiangnan cuisine. Jiangnan cuisine belongs to mid-range cuisine and was mainly distributed in Jiangsu and Zhejiang provinces in the south of the Yangtze River valley, including Huaiyang cuisine, Suzhou cuisine, Wuxi cuisine, Shanghai cuisine, Hangzhou cuisine, etc. Shanghai cuisine is a little bit sweet and particular about greasy and red sauce. Hangzhou cuisine is mainly fresh fish and vegetable dishes, and its taste is mainly salty and fresh. Li Qigui, through his good friend Zhao Renliang, Chief Chef of Kunlun Hotel, invited a chef of Shanghai cuisine. Apart from authentic Suzhou and Wuxi famous dishes, he mainly managed to improve Shanghai cuisine, making it more suitable for the tastes of northerners. For example, River Fresh Soup used crucian carp, river shrimp and river crab as the main ingredients, and after washing, frying and then cooking soup, it tasted extremely fresh.

Then, after Li Qigui arrived at Sunworld Dynasty Hotel, with the consent of the supervisor and the director of catering, he took about a year to reasonably adjust the catering structure of Sunworld Dynasty Hotel, dividing the space of the original cafeteria on the ground floor into two parts and adding Beijing-style Shandong cuisine featuring Roast Duck with Ganoderma Lucidum on one side and Hunan cuisine hall featuring spicy food on another side. The original Four Seasons Hall on the

在李启贵大师全力以赴的经营下，天伦王朝饭店的餐饮收入，从1998年他刚来时候的1200多万，到第二年就翻了一番，达到2400多万，2000年是3600多万，2001年达到4000多万，2002年到了5000多万，营业收入就这样头也不回地一路攀升上去了。

做五星级酒店，每年要有推广活动。饭店的规律是，在生意淡季的时候做推广，为的是招揽顾客。根据季节，要中西餐合理搭配。2000年，李启贵的第一个推广活动是什么呢？他起了个名字叫"六大名厨闹京都"。所谓"六大名厨闹京都"，就是邀请六位知名的大厨在这一年当中推出六个菜系。可是排兵布阵、逗引埋伏，大家都看总厨师长李启贵的令旗令箭。李启贵推出的第一个菜系是什么呢？是港式精品粤菜。第二个推闽菜的佛跳墙。第三个推杭州菜。第四个把聂厚忠大师请来推湘菜。第五个推京鲁菜、东北菜。第六个推上海菜。

菜系定完之后，要起个什么名头才好推广呀？大家凑在一起开始讨论。有的说叫"四季"，众人觉得太俗了，可一时又没什么好主意。饭店公关部经理潘高峰凑到李启贵跟前问："大师，你有什么想法？"李启贵想了想说："我别的不知道，我倒知道历史上有一个'五鼠闹东京'是不是？你看咱们请的人，都是全国最有名的高人，可称之为'六大名厨闹京都'。这些大师不但亲自出马亲自掌灶，还都给这次推广活动出高招儿想妙策。广州玫瑰园的名厨陈泽廷，粤菜一级棒！他建议咱们在粤菜里添加一些生啫类的客家菜肴。闽菜的'佛跳墙'强振涛来做，他是出身烹饪世家的闽菜大师。振涛师弟还给咱们出主意，建议咱们别光做'佛跳墙'，他说福建菜里边能和佛跳墙相媲美的，比如氽海蚌、红糟鸡、红糟鸭、泉州面线糊、海蛎窝等等，你看多丰富啊！"

一聊起这些，李启贵就打开了话匣子。他准备第三个推上海菜的名厨，是昆仑饭店总厨赵仁良推荐的名厨陈广跃，也是赵仁良的得意门生。杭州菜推的是"迷踪菜"的泰斗胡中英、王仁孝，老二位把杭州的杭椒、水芹、芦蒿直接引进到了天伦王朝饭店。做东北菜，李启贵找的是全国烹饪大赛评委、哈尔滨的名厨高峰，他推荐了自己的高徒、黑龙江省烹饪大师姜立滨。姜立滨拿手的是压锅菜汁浓味厚的得莫利炖鱼、熏腔骨和哈尔滨红肠等。阵容强大，高手如云。

事情就这么定下来了，开弓没有回头箭，说干就干。这"六大名厨闹京都"的推广活动持续了一年，声势之大，效果之好，在中国烹饪界中果然不同凡响。就连世界中国烹饪联合会会长兼中国烹饪协会会长张世尧、秘书长林则

second floor was also equally divided: the west side was renamed Shenghuaxuan Restaurant to operate excellent Hong Kong-style Cantonese cuisine and the east had become a gourmet world for Shanghai cuisine. The pattern of the four major cuisines was thus formed, which also laid the foundation for the development of the catering structure of the Sunworld Dynasty Hotel for more than ten years, making Sunworld Dynasty Hotel reach its heyday in catering management. All pays deserved gains. With Master Li Qigui's all-out operation, the catering income of Sunworld Dynasty Hotel doubled from more than RMB12 million in 1998 when he just arrived to more than RMB24 million in the following year, more than RMB36 million in 2000, more than RMB40 million in 2001 and more than RMB50 million in 2002. Thus, the operating income rose continuously without dropping.

To be a five-star hotel, there must be promotional activities every year. As a rule, the hotel made promotion during the off-season to attract customers. According to seasons, Chinese and Western dishes should be reasonably matched. What was Li Qigui's first promotion in 2000? He called the promotion Six Famous Chefs Competing in Beijing, inviting six famous chefs to launch six cuisines during the year. However, in terms of layout and design of the performance, everyone followed Chief Chef Li Qigui. What was the first cuisine introduced by Li Qigui? It was excellent Hong Kong-style Cantonese cuisine. The second was Steamed Abalone with Shark's Fin and Fish Maw in Broth, which was Fujian cuisine. The third was Hangzhou cuisine. The fourth was Hunan cuisine cooked by Master Nie Houzhong. The fifth was Beijing-style Shandong cuisine and Northeast cuisine. The sixth was Shanghai cuisine.

After the cuisines were decided, what title of promotion should be given? They gathered together to begin the discussion. Some said it could be called "Four Seasons", however most staff thought it too ordinary without any good ideas at that moment. Pan Gaofeng, Manager of the Public Relations Department of the Hotel, approached Li Qigui and asked, "Master, what do you think?" Li Qigui thought for a moment and said, "I don't know anything else, but I know the Invincible Constable in history, which meant five heros fighting for justness in Kaifeng (Capital City of Song Dynasty) in Chinese literally? As you see, we would invite most famous experts across the country, so we could call it Six Famous Chefs Competing in Beijing. These masters not only personally cooked, but also gave the promotion clever ideas. Chen Zeting, a famous chef in Guangzhou Rose Garden Restaurant, could cook first-class Cantonese cuisine. He suggested to add some Hakka dishes with raw gel to Cantonese cuisine. The Steamed Abalone with Shark's Fin and Fish Maw in Broth of Fujian cuisine was cooked by Qiang Zhentao, a master of Fujian cuisine from a culinary family, who suggested to cook other dishes, which were comparable to Steamed Abalone with Shark's Fin and Fish Maw in Broth in Fujian cuisine, such as Blanching Mussels, Braised Chicken with Rice Sauce, Braised Duck with Rice Sauce, Quanzhou Noodle Paste, and Oyster Nest. How rich they were!"

As soon as we came to this, Li Qigui talked without a stop. He was going to promote the famous chef Shanghai cuisine in the third position, who was Chen Guangyue, a famous chef recommended by Chef Zhao Renliang of the Kunlun Hotel, and was also Zhao Renliang's favorite apprentice. In terms of Hangzhou cuisine, he was going to promote Hu Zhongying and Wang Renxiao, the leading figures of Mixed Cuisine, who directly introduced Hangzhou's Hangzhou pepper, cress and Artemisia selengensis to Sunworld Dynasty Hotel. For Northeast cuisine, Li Qigui was looking for Gao Feng, a judge of the National Cooking Contest and a famous chef in Harbin, who recommended Jiang Libin, his senior apprentice and culinary master in Heilongjiang Province. Jiang Libin was good at Stewed

普，走到哪儿都说，做五星级酒店的推广活动，最成功、最有品位的就是天伦王朝李启贵大师做的，声誉做得很好，生意很兴隆，客人满意度也很高，天伦王朝的餐饮声名远播。

  2001年，李启贵陪同天伦王朝总裁吴晓燕、副总经理安学峰，接待了以法国大使为首的毛里求斯大使、乌拉圭大使、墨西哥大使，共同议定由天伦王朝派出名厨到毛里求斯做美食节活动，交流展示中国悠久的饮食文化。同年，李启贵应邀到比利时，考察了西方的西餐烹饪技法，并与华人一起烹制了中西合璧的菜肴。

  2002年，法国餐饮烹饪协会驻远东的主席古扎丽和中国驻法国大使，在天伦王朝饭店授予李启贵法国餐饮界的最高荣誉"蓝带奖"。

  2004年，李启贵大师在北京天伦王朝饭店接待了由世界烹饪联合会会长张世尧陪同前来考察的西班牙大使，详细地介绍了天伦王朝饭店的餐饮结构、风味构成及特色，无形中传播了中华悠久的饮食文化。

  不要以为管着上百名高级厨师的李启贵就远离灶台了。尽管他是酒店的总厨师长，管理上需要操心的事很多，但是他始终认为：身教重于言教。如果总厨都不上灶炒菜，那就失去了当总厨的意义。如果总厨不上灶，这一个厨房的风味把握就可能失控了。

  在国外，讲究米其林星级评定。他们规定，大厨不上灶，摘掉一颗星。第二次查，大厨还不上灶，摘掉两颗星，直至把星给摘完了。李启贵经常到国外考察，对此深有感触。作为总厨师长，李启贵虽然不能在所辖的38家五星级酒店都做到亲自操作，但在北京天伦王朝饭店本部，所有的重要宴会，前三道菜都是李启贵亲自操作。

  一次，上海锦江饭店举办挪威首相欢迎宴会，特地把李启贵大师从北京接到了上海。李启贵亲自主理的是他的拿手名菜"风生水起"，28桌嘉宾现场共睹李启贵大师烹饪的风采。"风生水起"的原料选用新鲜的三文鱼、海蜇、生菜、青红柿、西芹、葱丝、胡萝卜、白萝卜、黄瓜、洋葱头，以及XO酱、生抽、精盐、芝麻油、蚝油、绿芥末、花生、腰果、芝麻、柠檬汁等调味料。先把三文鱼切薄片，海蜇、生菜等各种青菜切丝，芝麻焙香，花生、腰果烤香切碎。绿芥末加适量生抽澥开，将生菜丝、海蜇丝放在盘子中间垫底。切好的三文鱼整齐地码在上面，周围按照色彩协调的原则码放好各种菜丝。花生等小料放在大盘边上。吃的时候，将各种菜丝放在鱼片上，加入各种调料拌匀即成。

Fish at Demoli, Smoked Bones and Harbin Sausage with thick and mellow pot juice. The lineup was strong with many experts.

All were settled and what's done couldn't be undone. The promotion of Six Famous Chefs Competing in Beijing lasted for one year, with great momentum and good results. It was indeed extraordinary in the Chinese culinary circle. Even Zhang Shiyao, President of the World Association of Chinese Cuisine and President of the China Cuisine Association, and Lin Zepu, Secretary General of the latter, regarded it as the most successful and elegant promotion activities for five-star hotels, which were done by Master Li Qigui from Sunworld Dynasty Hotel. With good reputation and prosperous business, Sunworld Dynasty Hotel received higher satisfaction from guests, which further made it widely known.

In 2001, Li Qigui accompanied Wu Xiaoyan, President of Sunworld Dynasty Hotel, and An Xuefeng, its Deputy General Manager, to receive the ambassadors of Mauritius, Uruguay and Mexico headed by the French ambassador. They jointly agreed that Sunworld Dynasty Hotel would send famous chefs to Mauritius to prepare food festival activities so as to exchange and display Chinese long-standing food culture. In the same year, Li Qigui was invited to Belgium to inspect Western cooking techniques and cook Chinese and Western dishes together with overseas Chinese.

In 2002, Guzari, Chairperson of the French Catering and Cooking Association in the Far East, and the Chinese ambassador in France awarded Li Qigui the "Blue Ribbon Badge", the highest honor in French catering industry, at the Sunworld Dynasty Hotel.

In 2004, Master Li Qigui received the Spanish ambassador accompanied by Zhang Shiyao, President of the World Cooking Federation, at Sunworld Dynasty Hotel Beijing. He gave a detailed introduction to the catering structure, flavor composition and characteristics of Sunworld Dynasty Hotel, which virtually spread Chinese long-standing catering culture.

Although Li Qigui was in charge of hundreds of senior chefs, he still went around the cooking bench. As the chief chef of the hotel to manage many things, he always thought, "Example is better than precept. If a chef does not cook, it will lose the duties of being a chef. If a chef does not cook, the flavor of this kitchen may be out of control."

In foreign countries, Michelin star rating was emphasized. It's stipulated that a chef should be degraded by taking off one star if he doesn't cook. On the second check, the chef should be degraded by taking off another two stars if he doesn't cook until all stars are taken off. Li Qigui often goes abroad for inspection and is deeply touched by this. As the chief chef, although Li Qigui cannot operate in person in all the 38 five-star hotels under his jurisdiction, Li Qigui personally cooked the first three dishes for all the important banquets in the headquarters of Sunworld Dynasty Hotel.

On one occasion, Shanghai Jinjiang Hotel was going to hold a welcome banquet for the Norwegian Prime Minister and specially invited Master Li Qigui from Beijing to Shanghai. Li Qigui personally managed his famous dish Feng Sheng Shui Qi (meaning thriving). 28 tables of guests saw the elegant demeanour of Master Li Qigui's cooking. Feng Sheng Shui Qi used such main ingredients as fresh salmon, jellyfish, lettuce, green and red sweet pepper, celery, shredded scallion, carrot, radish, cucumber and scallion, and seasonings like XO sauce, raw soy sauce, refined salt, sesame oil, oyster sauce, green mustard, peanut, cashew nut, sesame, lemon juice and others. First, slice salmon, shred jellyfish, lettuce and other green vegetables, roast sesame, peanut and cashew nuts, and chop them up. Add a proper amount of green mustard to thin down, and place lettuce shreds and jellyfish shreds in

此菜三文鱼肉质鲜嫩，配料爽口，滋味香辛，是宴席开胃的至味佳品。在天伦王朝饭店，凡是最重要的宴会前三道菜，都是李总厨亲自操作，第一道菜"风生水起"，一捞福气，二捞财气，三捞运气。第二道菜就是他发明的中华八珍宝鼎，在宴会厅里推上宝鼎现场制作。第三道菜是芙蓉蟹肉雪蛤，也是现场用砂锅炒，然后让服务员一人一份现场给贵宾们分餐。李启贵大师的厨艺如何，现场所有嘉宾大快朵颐的表情最具感染力，最有说服力。

在李大师现场制作的这些精美菜肴当中，有一道让人叹为观止的菜品。这道菜竟然还获得了国家专利局颁发的发明专利，发明人正是李启贵大师本人。这在中国烹饪界简直是凤毛麟角，别开天地。究竟是什么发明，让国家专利局的评审专家们对一位厨师刮目相看呢？

the middle of the plate as the bottom. Stack the cut salmon neatly on it, and all kinds of shreds around according to the principle of color coordination. Place peanuts and other seasonings on the edge of the plate. When eating, put all kinds of shredded vegetables on the fish fillets, add all kinds of seasonings and mix well. The salmon of this dish was fresh and tender in meat, refreshing in ingredients and spicy in taste. It's an appetizing food for banquets. In Sunworld Dynasty Hotel, Chef Li personally cooked the first three dishes of all most important banquets. The first dish was Feng Sheng Shui Qi, meaning for good fortune, wealth and luck. The second dish was the Chinese Eight Treasures Tripod invented by him, which was cooked on the spot with the cauldron pushed to the banquet hall. The third dish was Chinese Forest Frog & Crab Meat with Lotus, which was also fried in a casserole on the spot, and then the waiter would present the meal to the distinguished guests by parts one by one. Master Li Qigui's cooking skills were fully demonstrated by the most infectious and convincing expression of all the guests eating at the scene.

Among these exquisite dishes made by Master Li on the spot, there was an amazing one. This one even won an invention patent issued by the National Patent Office of the China National Intellectual Property Administration. The inventor was Master Li Qigui himself. It was rare in the Chinese culinary circle to perform something different. What kind of invention makes the evaluation experts of the National Patent Office hold him in high esteem?

中国十佳烹饪大师李启贵在主持天华国际集团烹饪大赛。

Li Qigui, one of China's top 10 culinary masters, presided over the cooking competition of Tianhua International Group.

2000年2月，李启贵大师参加中国大陆、香港、澳门、台湾四位大师迎千禧节目，大师们各施绝技，李启贵大师烹制名菜"九龙八珍宝鼎宴"。

In February 2000, Li Qigui participated in the Chinese, Hong Kong, Macau and Taiwan culinary program with three other masters, and Li Qigui cooked the famous dish—Chinese Jiulong Eight Delicacies Banquet.

2000年，李启贵大师在天伦王朝为挪威大使亲自操作广东名菜"风生水起"。

In 2000, Li Qigui worked with Norwegian chefs at Sunworld Dynasty Hotel to make the famous Cantonese dish—Fengsheng Shuiqi.

2000年7月，中国驻美国30年大使柴泽民先生在天伦王朝为行政总厨李启贵大师题词。

In July 2000, Mr. Chai Zemin, China's ambassador to the United States for 30 years, presented an inscription to Executive Chef Li Qigui at Sunworld Dynasty Hotel.

2000年10月26日，李启贵大师应邀到德国考察精品酒店无人服务流程。

On October 26, 2000, Li Qigui was invited to Germany to study the unmanned service process of boutique hotels.

李启贵大师任第四届全国大赛评委。
Li Qigui served as the judge of the 4th World Championship of Chinese Cuisine.

2005年，王义均与李启贵师徒两人在人民大会堂，于第五届全国高新技术比赛颁奖大会的闭幕宴会上开心交谈。
In 2005, the master and the disciple chatted happily at the closing banquet in the Great Hall of the People.

李启贵大师与恩师周子杰、徒弟贾河武一起向中烹协领导介绍天伦王朝满汉全席。
Li Qigui, together with his mentor Zhou Zijie and apprentice Jia Hewu, introduced Sunworld Dynasty Manchu–Han Imperial Feast to the Chinese culinary leaders.

李启贵大师烹制的中华八珍宝鼎色香味形俱佳，深受欧洲美食家赞誉。
Li Qigui's Chinese Eight Delicacies Banquet were highly praised by European gourmets for their excellent color, aroma and style.

第四届中国烹饪世界大赛颁奖礼。
Award Ceremony of the 4th World Championship of Chinese Cuisine.

李启贵应聘为"第七届中国烹饪世界大赛"国际评委颁奖的"大宴席"席间。
The Grand Banquet where Li Qigui was appointed to present the award for the 7th World Championship of Chinese Cuisine.

于京贵八珍宝鼎宴认定会上。
Recognition Meeting for the Beijing Eight Delicacies Banquet.

出席中国烹饪协会第二届监事会第一次会议。
The First Meeting of the Second Supervisory Board of China Cuisine Association.

2020年，李启贵大师出席丰泽园建店90周年典礼。
In 2020, Li Qigui attended the 90th anniversary ceremony of the establishment of Fengzeyuan.

第六回

Chapter Six

妙用灵芝 绝品迷住四海客
情系宝鼎 八珍融入五千年

1998年夏天，秦皇岛召开了一个全国烹饪界备受瞩目的会议，李启贵大师作为北京地区的唯一代表出席了这次会议。这次会议的中心议题是，研究起草制定"中国烹饪大师""中国烹饪名师"称号的评选条件和评选标准。在这次会议上，李启贵和他的老师、中国烹饪协会副秘书长李正权先生同住一个房间，两人聊得很投机，经常能讨论到夜里三四点。聊着聊着，李启贵就把心里一个多年的想法跟李先生说了出来。

李启贵当时说："李老，我这儿有一想法，我想研究一个宝鼎的菜。过去老泰丰楼有道名菜叫'一品锅'，我一直想恢复和完善这道名菜，可是研究了半天，也没有合适的容器，有没有可能用鼎这个容器代替原来的铜锅呢？"

李正权点了点头说："未尝不可一试，你是怎么想的呢？"李启贵一听有门儿，就把自己几年来的想法都倒了出来。他说，想结合中国5000年的饮食文化，再现中国的烹饪历史。鼎，最早是人类做饭做菜的食器，后来发展成为礼器，成为权力的象征。我想把鼎再恢复它的本来功用，作为美食器具，一是对身体有益；二是庄重大气。

李启贵之所以会产生这个看似异想天开的念头，还是源于他的师父王义均大师的一次家宴。1986年李启贵从国际大赛上拿回金牌不久，王义均叫李启贵到家里吃饭，还说："我给你做一个新的菜，你保管没吃过。"到了师父家，王义均大师端出一个大石锅来，给爱徒做了一个海鲜汤，里边放的鸡蓉、海白菜、海胆之类的东西。李启贵盛出来一尝，嗬！还真鲜美！

这个石锅海鲜汤给李启贵留下了极其深刻的印象，在心里埋下了一颗创新的种子。从这儿以后，这个想法不停地在李启贵的脑子转悠。他业余时间爱读书，围绕中国的钟鸣鼎食文化，他翻阅了大量书籍来考证。这次跟李正权先生深度探讨，这个想法逐渐清晰定型了。

术业有专攻，做玉鼎，李启贵自然想到了他的雕刻老师崔技良。提起崔大师，李启贵赞不绝口："我这老师是北京雕刻厂的厂长，他为人甭提多好了，教我雕一样东西，那个认真劲儿让我都感动。有时候他到我家教我，一雕雕到深夜，然后自己骑自行车回家，就喝口水，饭也不吃。"

Fabulous Dishes with Ganoderma Lucidum Enchanting Tourists across the World
Eight Treasures in Tripod Integrating the History of 5,000 Years

In the summer of 1998, a high-profile conference in the national culinary circle was held in Qinhuangdao. Master Li Qigui, as the only representative in Beijing, attended the conference. The central topic of this meeting was to study, draft and formulate the selection conditions and criteria for the titles of "Chinese Cooking Master" and "Famous Chinese Cooking Chefs". At this meeting, Li Qigui and his instructor Mr. Li Zhengquan, Deputy Secretary-General of the China Cuisine Association, shared a room. With much in common, they often discussed until 3 or 4 o'clock at night. While chatting, Li Qigui told Mr. Li an idea in his mind for many years.

Li Qigui said at that time, "Senior Li, I have an idea here. I want to study a dish with a tripod. In the past, there was a famous dish called 'First Class Pot' in the old Taifeng Restaurant. I have always wanted to restore and improve this famous dish, but after studying for some time, there was no suitable container. Is it possible to replace the original copper pot with a tripod?"

Li Zhengquan nodded and said, "It is not impossible to have a try. What do you think?" As soon as Li Qigui heard it possible, he threw out all his thoughts over the past few years. He said that he wanted to combine Chinese 5000-year food culture and reproduce Chinese cooking history. Tripod, originally a food vessel for human beings to cook, later developed into a ritual vessel and became a symbol of power. I wanted to restore the tripod to its original function as a gourmet utensil. First, it was beneficial to the human body. The second, it was solemn and generous.

Li Qigui had this seemingly whimsical idea due to a family dinner hosted by his master, Master Wang Yijun. Shortly after Li Qigui won the gold medal from an international competition in 1986, Wang Yijun asked Li Qigui to dinner at home and said, "I'll cook you a new dish and you must have never had it." When he arrived at Master Wang Yijun's house, Master Wang Yijun brought out a large stone pot and cooked a seafood soup for his beloved apprentice, which contained chicken paste, sea cabbage, sea urchin and other things. Li Qigui took it out and had a taste. Wow! It's delicious!

This stone pot seafood soup left a deep impression on Li Qigui and planted a seed of innovation in his mind. Since then, this idea had been wandering around in his mind. He loved reading in spare time. Around Chinese food culture of lining up tripods of eatables and playing musical performance during a meal, he had read a large number of books to verify it. His in-depth discussion with Mr. Li Zhengquan made that idea gradually clear and finalized.

A specialist only masters his own field. In terms of making a tripod, naturally, Li Qigui thought of his sculpture instructor Cui Jiliang. Speaking of Master Cui, Li Qigui praised profusely, "My instructor is Director of Beijing Carving Factory. He is pretty kind. I was often moved by his seriousness. Sometimes he came to my house to teach me carving until late at night, and then rode his bicycle home, only drinking some water but not letting us entertain him for dinner."

崔老师听说李启贵要用鼎烹饪，连声叫好，说雕刻玉鼎的事没问题，只是这个纹饰挺麻烦的，鼎身上得有饕餮纹。崔大师说："吃大餐，就是要如饕如餮。"事情敲定下来，崔大师就开始设计方案，为了强调专属性，还在鼎的底部和两耳都雕上了"中国烹饪大师李启贵"的篆文字样，在北京雕刻厂完成了第一次雕刻。中国玉石鼎采用的是"鼎中鼎"的设计方案，即鼎的上面有锅，锅的下面有火，锅和底座之间有夹层支着，烹炒煮炖，可以随心所欲。样品研制成功后，批量生产跟了上来，最多的一次，李启贵定制了38个玉石鼎，足足运了一车加长的130卡车。"工欲善其事，必先利其器"，鼎做好了，就成功了一半，下面就是李启贵大师如何讲好宝鼎故事，如何调和鼎鼐烹小鲜了。

制作中华八珍宝鼎的原料比一般菜肴要复杂得多。主要包括水发鱼翅、鲍鱼、鱼唇、水发海参、水发裙边、鱼骨、水发鱼肚、干贝、大虾肉、虎掌菌、羊肚菌、松茸菌、猴头蘑、油菜心、清汤、奶汤、精盐、胡椒粉、葱姜油、熏干贝的原汤、鸡蛋清、水淀粉、食用油等。具体制作方法是，鱼翅用清汤煨透入味，鲍鱼、鱼唇、海参、鱼肚、裙边均改刀成片状。鱼骨切成小长方条，大虾肉片成片儿，加蛋清、淀粉浆好滑熟。干贝去筋蒸透保留原汤。虎掌菌、猴头蘑改刀成片状，连同羊肚菌、松茸片用清汤煨好。小油菜心焯透入上味，鼎内放入奶汤、清汤，加入主配料、精盐、料酒、葱姜油、胡椒粉、干贝原汤，旺火烧开，下入油菜心和鱼翅即成。此菜集众多珍贵原料于一鼎，汤鲜味浓，回味无穷，闻之清香四溢，食之软嫩滑爽，观之色彩艳丽，用之营养丰富。它的烹饪方法不同于传统的酒焖，而是采用烹炒兑原味浓汤的办法，使原料更加有形，口味更加细腻。在调理原则上有别于南方的滋补汤。南补汤的调理概念是去热、除湿、散闷、适补的要求；中华八珍宝鼎则改为适应北方传统口味，其调理概念是去燥、润干、清热、适补。

作为天伦王朝饭店的"镇店第一名菜"，必须赋予它足够的历史文化内涵。笔者曾在旧作《九鼎颂》中写道："结绳之记，数唯九大；奉夏之祀，器唯鼎尊。铸九牧之金，铭一统之始。禹王登极，鼎定以昭日月；周室式微，金销而偃春秋。毛遂一言，使楚重于九鼎；庄王试问，在德迁于三代。伊尹施政，理朝堂如调鼎鼐；老子箴言，治大国如烹小鲜。国家重器，物中匡圣；社稷重臣，人中宝鼎。享祭祝颂，生民乐而未央；吐哺归心，士林望而仰止。"其实从中国的新石器时期开始，人们就有意识地把食物放在削磨过的石板上，利用灼热的太阳把食物烤热晒焦。雷电交加，引燃了大火，烤熟的食物又沾到

When Mr. Cui heard that Li Qigui was going to cook with a tripod, he cheered repeatedly, saying that there was no problem in carving a jade tripod, but the decoration was quite troublesome for the tripod had to have Taotie patterns. Master Cui said, "To have a big meal is to be like a Taotie to enjoy the meal greatly." When the matter was finalized, Master Cui began to design the plan. In order to emphasize the specificity, the seal character "Chinese Culinary Master Li Qigui" was carved on the bottom and both ears of the tripod, and the first carving was completed in Beijing Carving Factory. Chinese jade tripod adopted the design scheme of "Tripod in Tripod", namely a pan on the top of the tripod, a fire under the pan, and an interlayer between the pan and the base, which can be used for cooking, frying, boiling and stewing at will. After the sample was successfully developed, mass production followed. At one time, Li Qigui customized 38 jade tripods, being fully transported by an extended truck No. 130. If a worker wanted to do a good job, he must first sharpen his tools. Once the tripod was ready, it's half successful. The following is how Master Li Qigui told the story of the tripod and how to cook various dishes with the tripod.

The main ingredients for cooking the Chinese Eight Treasures Tripod were much more complicated than ordinary dishes, mainly including water-fat shark's fins, abalones, shark's lips, water-fat sea cucumbers, water-fat turtle rim, fishbone, water-fat fish maw, dried scallops, prawn meat, sarcodon aspratus, morel, tricholoma matsutake, hericium erinaceus, rape, clear soup, milk soup, refined salt, pepper, fried scallion-ginger oil, raw soup of smoked dried scallops, egg white, water starch, edible oil, etc. The specific preparation method comprised the following steps: simmering shark's fins with clear soup thoroughly, and cutting abalones, shark's lips, sea cucumbers, fish maw and turtle rim into slices. Cut the fish bones into small rectangular strips, cut the prawn meat into slices, and add egg white and the starch paste to make them smooth. Remove tendons and steam dried scallops to keep the original soup. Cut sarcodon aspratus and hericium erinaceus into slices, and simmer with clear soup together with morel and tricholoma matsutake slices. Blanch rape, add seasonings, add milk soup and clear soup into the tripod, add the main ingredients, ingredients, refined salt, cooking wine, fried scallion-ginger oil, pepper and raw dried scallop soup, boil over strong fire, and add rape and shark's fins. This dish combined many precious main ingredients in one tripod. The soup had strong delicate flavor, leading a person to endless aftertastes. It smelled fragrant, tasted soft, tender and smooth, looked bright color and had rich nutrition. Its Steps was different from the traditional stewing with wine, but adopted the stir-frying with original thick soup to make the main ingredients more shaped and the taste more delicate. In principle of conditioning, it was different from the nourishing soup in the south. The conditioning concept of the nourishing soup in the south was to remove heat, dehumidify, reduce boredom and supplement properly. The Chinese Eight Treasures Tripod was suitable to the traditional tastes of the north. Its conditioning concept was to remove fret, moisten, reduce heat and supplement properly.

As the "Most Famous Dish" of Sunworld Dynasty Hotel, it must be endowed with sufficient historical and cultural connotation. The author once wrote in the previous work *Ode to the Nine Tripods*, "In ancient times when there was no writing, people tied knots on ropes to keep records and took nine as the largest number. In the Xia Dynasty, the tripod was regarded as the most important ritual vessel. The King of Xia divided the country into nine states, each casting a bronze tripod, which heralded the unification of the country. At the end of the Zhou Dynasty, the royal family declined, feudal lords vied for the throne and the nine big tripods disappeared. Mao Sui went to Chu State to fulfill a diplomatic mission and asked Chu State to send troops to save Zhao State so as to keep the country safe. King Zhuang of Chu once asked Wang Sunman, the official of the Zhou Dynasty,

了海水的咸味。李启贵认为，烧焦的食物是烹的开始，沾到海水有了滋味，这是调的开始。这就是烹调的起源。

在天伦王朝饭店推出中华八珍宝鼎宴的时候，李启贵曾经接待了一次172个国家大使参加的盛大宴会。他在宴会厅现场烹制宝鼎宴的当中，边做边说："鼎是一种古代礼器，也是权力的象征，它象征着世界各国人民的团结，也象征着世界各国人民经济事业的飞速发展。"大使们闻听纷纷点头赞许。后来，李启贵又带着宝鼎到美国表演，到荷兰展示，到新加坡交流烹饪，这番精彩的诠释，众人听了无不折服。

当然，中华八珍宝鼎也为李启贵赢得了更多的荣誉。2000年首届中国厨师节在杭州举行，李启贵被评为"中国十佳烹饪大师"。为此，天伦王朝饭店专门给李启贵大师出了首日封，以示庆贺。

同样是千禧之年，在中央电视台举办的中国大陆、香港、澳门、台湾四位大师迎千禧的节目中，代表香港的是李波大师，代表台湾的是许长云大师，代表澳门的是黄永距大师。代表大陆出场的李启贵大师，现场制作的是"九龙八珍宝鼎"，上面宝鼎里的美味现场烹制，下边用南瓜雕了9条龙，连嘉宾主持侯耀文先生都赞叹不已。侯耀文先生在一旁观看李启贵大师烹制宝鼎佳肴的新闻照片还刊登在2000年2月13日的《北京晚报》上。

最让李启贵难忘的是，世纪之交，天伦王朝饭店在室内广场举办了一台"丝绸之路"晚会，邀请了10年来的模特冠军来此走秀，还特别聘请了一个青春靓丽的女子装扮成公主的模样，可是器宇轩昂的国王去哪儿找呢？晚会组织者一眼看中了方面大耳、身材魁伟、相貌堂堂的李启贵大师，他正在宴会厅现场烹制他的中华八珍宝鼎。赶忙让他换了总厨服装，解下围裙，穿上国王的袍服，戴上金灿灿的王冠，粘上浓黑的八字胡须，右手挂着一根镶着红宝石的王权手杖。大家一看，果然是一派帝王气象！都笑着说："这个国王最像样！还是免费的！"

中华八珍宝鼎宴一经推出，好评如潮，经久不息。

2000年7月，中国驻美国大使柴泽民因外事活动考察天伦王朝的餐饮服务工作，为李启贵大师烹制的八珍鼎宴题词："树天伦品牌，创一流业绩"。2002年，全国政协副主席孙孚凌在烹坛泰斗王义均的陪同下，考察了天伦王朝饭店的餐饮服务工作。李启贵当着师父王义均的面，现场为一行贵宾表演烹制了中华八珍宝鼎、炒燕窝和北京灵芝烤鸭。他精湛的厨艺，得到了孙孚凌先生

about the weight of the nine tripods. In fact, he coveted the situation of the Zhou Dynasty. Yi Yin, the prime minister of the Shang Dynasty, governed the country like seasoning ingredients in the tripod. Lao Zi (Li Er) proposed in Tao Te Ching that governing a large country should be like cooking delicious dishes. Tripod is an important utensil of the country, and the ministers who run the country are treasures like tripod among the people. Solemnly offering sacrifices to the sages with a tripod and regarding the sages as treasures, the common people will live good lives forever. Like the Duke of Zhou, who had three breaks during a meal to receive guests, was eager for talents. Only then will the talented and wise men in the world be committed to the country and full of respect for the rulers." In fact, since the Neolithic Age in China, people had consciously placed food on polished stone slabs, which was crisp by the scorching sun. Thunder and lightning ignited the fire, and the baked food stained with the salty taste of sea water. Li Qigui believed that scorching was the beginning of cooking, and it tasted well with seawater, which was the beginning of seasoning. This was the origin of cooking.

When Sunworld Dynasty Hotel launched the Chinese Eight Treasures Tripod Banquet, with which Li Qigui once received a grand banquet attended by ambassadors from 172 countries. While cooking the tripod banquet in the banquet hall, he said, "Tripod is an ancient ritual vessel and a symbol of power. It symbolizes the unity of the people and the rapid development of the economic undertakings of all countries in the world." The ambassadors nodded approvingly. Later, Li Qigui took tripod to perform in the United States, to show in Holland, and to exchange cooking in Singapore. All the people were impressed by his wonderful interpretation.

Of course, the Chinese Eight Treasures Tripod Banquet had also won more honors for Li Qigui. In 2000, the first Chinese Chefs' Festival was held in Hangzhou, and Li Qigui was named China's Top Ten Cooking Masters. For this reason, Sunworld Dynasty Hotel specially issued a special first day cover to Master Li Qigui to celebrate.

In 2000, the year of the millennium, CCTV held a program to welcome the millennium by four masters from Chinese mainland, Hong Kong, Macao and Taiwan, among whom Master Li Bo represented Hong Kong, Master Xu Changyun represented Taiwan, Master Huang Yongji represented Macao, and Master Li Qigui represented Chinese mainland, who made the Nine Dragon-shaped Eight Treasures Tripod on the spot. The delicious food in the tripod was cooked on the spot, and nine dragons were carved with pumpkins on the bottom. Even the guest host, Mr. Hou Yaowen, was amazed. The news photos that Mr. Hou Yaowen was watching Master Li Qigui cooking delicacies with a tripod was also published in the Beijing Evening News on February 13, 2000.

What impressed Li Qigui most was that at the turn of the century, Sunworld Dynasty Hotel held a "Silk Road" party in an indoor square, inviting the model champion of the past 10 years to give a show there, and specially hiring a young beautiful woman to dress up as a princess. But where can they find the grand king? The organizer of the party took a fancy to Master Li Qigui, who was square-faced, big-eared, tall and handsome, and was cooking the Chinese Eight Treasures Tripod on the spot in the banquet hall. He was asked to quickly change the chef's clothes, take off his apron, put on the king's robe and the golden crown, wear a thick black moustache, and lean on a royal cane inlaid with rubies in his right hand. What an imperial style! Everyone smiled and said, "He is the best role as well as free!"

Once the Chinese Eight Treasures Tripod was launched, it was well received forever.

In July 2000, Chinese Ambassador in the United States Chai Zemin inspected the catering service of Sunworld Dynasty Hotel for foreign affairs activities and wrote an inscription for the Chinese Eight

和师父王义均的充分肯定。

2007，年荷兰格罗宁根市一个古城堡花园新开业了一家"楼外楼"中餐厅，是华侨吴洪钢先生开的。他久闻李启贵的大名，在开业之际，特地请李启贵大师来到酒楼，现场烹制中华八珍宝鼎，格罗宁根市的市长和中国驻荷兰的领事都出席了活动，品尝了李启贵大师的中华八珍宝鼎。

2008年，奥运会开幕前夕，李启贵现场烹制中华八珍宝鼎的大幅镜头出现在中国的各个机场的电子屏幕上，旁边两行大字："烹鼎庆奥运，北京欢迎你"。

在中美元首会晤前夕，2012年2月11号，李启贵大师到美国纽约展示交流中国厨艺。他在纽约一家电视台的节目中现场表演制作雪花龙须面和中华八珍宝鼎。世界中国烹饪联合会会长出席了揭鼎仪式，莫天成、旅游局局长罗伯特到现场品尝了宝鼎宴。让李启贵觉得有趣而且记忆犹新的是，八珍宝鼎的烹制都是现在大鼎里做好，再给每位宾客分到面前的小鼎里，没想到汤勺还没上呢，罗伯特局长已经迫不及待地端着鼎喝上了。

中华八珍宝鼎制成以后，怎么想起申请专利了呢？李启贵说："我有一个徒弟叫关志群，也是天伦王朝饭店的一位厨师长。八珍宝鼎研制出来以后，他跟我说，师父您这个应该申请一个专利，要不然将来这怎么算啊？您是不是应该考虑这个问题？我说这申请需要哪些步骤啊？咱们做餐饮的，也不懂这个呀。关志群说，咱们这儿有个粤菜厨师，他们家有在专利局工作的，请他帮着问问呗。就这样我就到专利局去办申请了。1999年下半年，我的专利证书就拿到了。"

虽然申请了个人专利，但李启贵仍然不计任何报酬地在世界各地推广八珍鼎宴。为了弘扬中国的饮食文化，李启贵大师把传授一身绝艺当成普及中华饮食文明的路径和责任，走遍祖国的东西南北，身教言传。南边，2001年3月，他曾经到江苏的镇江宾馆，现场传授烹饪技艺，亲自做菜，亲自做美食推广。西边，1995年7月至10月，他和师父王义均一起到新疆的绿洲宾馆，为维吾尔族同胞五十多个人的烹饪班授课。东面，2001年，他和王义均大师、世界中国烹饪联合会秘书长林德普到烟台认定鲁菜之乡。北面，2003年，李启贵来到海拉尔做美食节，亲自做菜，烹制中华鼎宴，教给当地厨师很多菜品。

Treasures Tripod Banquet cooked by Master Li Qigui, "Build the Brand for Sunworld Dynasty Hotel and Create First-class Performance". In 2002, Sun Fuling, vice-chairman of the National Committee of the Chinese People's Political Consultative Conference, accompanied by Wang Yijun, a leading cook, inspected the catering service of Sunworld Dynasty Hotel. In front of Master Wang Yijun, Li Qigui performed and cooked the Chinese Eight Treasures Tripod, Fried Bird's Nest and Beijing Roast Duck with Ganoderma Lucidum for a delegation of distinguished guests. His exquisite cooking skills had been fully affirmed by Mr. Sun Fuling and Master Wang Yijun.

In 2007, an ancient castle garden in Groningen, Netherlands, opened a new Chinese restaurant "Building Beyond Building", which was operated by overseas Chinese Mr. Wu Honggang. He has heard of Li Qigui's famous name for a long time. When he started the operation of the restaurant, he specially invited Master Li Qigui to the restaurant to cook the Chinese Eight Treasures Tripod on the spot. The mayor of Groningen City and the Chinese Consul in Holland all attended the event and tasted Master Li Qigui's Chinese Eight Treasures Tripod.

On the eve of the opening of the 2008 Olympics in Beijing, a large-scale scene of Li Qigui cooking the Chinese Eight Treasures Tripod appeared on the electronic screens of various airports in China, with two lines of large characters beside it, "Cooking the Tripod to Celebrate the Olympics and Welcome to Beijing".

Before the Sino-US Leader Summit, Master Li Qigui went to New York to show and exchange Chinese cooking skills on February 11, 2012. In a TV program in New York, he performed on-site making of snowflake Longxu Noodles and cooking of Chinese Eight Treasures Tripod. The president of the World Association of Chinese Cuisine (WACC) attended the ceremony. Mo Tiancheng, and Robert, Director of the Tourism Bureau, went to the scene to taste the banquet. What made Li Qigui felt interesting and fresh in his memory was that the cooking of the Chinese Eight Treasures Tripod was well done in the big tripod first, and then each guest received a part in the small tripod in front of each. Unexpectedly, Director Robert couldn't wait any more to drink the soup in the small tripod before the spoon was provided.

Asking the reason for applying for a patent when the tripod for the dish Chinese Eight Treasures Tripod was made, Li Qigui said, "I have an apprentice named Guan Zhiqun, who is also a chief chef of Sunworld Dynasty Hotel. After the tripod is developed, he suggests me to apply for a patent to have its ownership. I asked the steps required for the application and he knew little about that as a layman. Guan Zhiqun knew that a Cantonese chef worked in the restaurant, whose family member worked in the Patent Office. Then we asked him to give a help. As a result, I went to the Patent Office to apply and I got my patent certificate in the second half of 1999."

Although he applied for a personal patent, Li Qigui still promoted the Chinese Eight Treasures Tripod Banquet all over the world without any compensation. In order to carry forward Chinese food culture, Master Li Qigui took imparting his unique skills as the path and responsibility to popularize Chinese food civilization. He traveled all over the country, teaching by precept and example. In the south, he went to Zhenjiang Hotel in Jiangsu Province in March 2001 to teach cooking skills on the spot and to promote dishes in person. In the west, he and Master Wang Yijun went to the Green Land Hotel in Xinjiang from July to October 1995 to teach cooking classes for more than 50 Uighur compatriots. In the east, he and Master Wang Yijun and Lin Depu, Secretary General of the World Federation of Chinese Cuisine, went to Yantai in 2001 to identify the hometown of Shandong cuisine. In the north, Li Qigui came to Hailar to cook on the food festival, personally cooking dishes and the Chinese Eight Treasures Tripod Banquet, teaching local chefs many dishes.

除了身价不菲的中华八珍宝鼎，李启贵大师发明的"灵芝烤鸭"也是天伦王朝饭店八珍宝鼎宴里的一道重要菜肴。"我经营的店里边都有烤鸭，全聚德的名菜鸭包翅就是蔡启厚大师亲自传授给我的。众所周知，灵芝有益寿延年、净血养生的功效，我在日本大阪工作的时候，就给他们炮制过灵芝高粱酒，大受欢迎。到了天伦王朝，我就把这灵芝和烤鸭结合起来了，融为灵芝饼、灵芝酱、灵芝鸭，把这鸭子用灵芝腌过。你看别处吃烤鸭，他那酱里边没有灵芝。他那个饼里边没有灵芝，咱这饼里有灵芝孢子粉，然后这烤鸭又拿灵芝腌过。"

"中华八珍宝鼎宴"里的第一道菜是中华宝鼎，第二道菜是葱烧万寿参，第三道菜就北京灵芝烤鸭，形成"三足鼎立"。北京灵芝烤鸭的原料包括宰好的瘦肉型填鸭、天然灵芝、饴糖水、葱段、甜面酱、黄瓜条、蒜泥、白糖等。灵芝烤鸭的技术主要有宰烫、开生制坯、灵芝的泡发加工、烤制、片鸭五道工序。灵芝烤鸭与其他烤鸭的不同之处在于烤制前、"堵塞"后、灌入灵芝和灵芝汤，随即挂色进行烤制。炉温一般保持在230℃至250℃，而且要随时调整火力。烤好后取出灵芝用以点缀，并用灵芝粉制成灵芝酱和灵芝饼，连同葱白、黄瓜条、蒜泥、白糖等随烤鸭一起上桌即成。这道菜的特点是皮脆肉嫩，香气扑鼻，有补肺益胃安神的作用。

鼎盛时期的天伦王朝餐饮，在李启贵大师的运筹中，真如那道他挖掘恢复的唐宋贺岁名菜一样"风生水起"。但是，如何把中华饮食文化更好地传承下去？李启贵比别人想得多些，做得则更多。

庄子在《养生主》一篇中说，"指穷于为薪，火传也，不知其尽也。""指"古通"脂"。古代没有发明蜡烛的时候，用脂肪裹薪点燃照明，谓之"烛薪"。这句话的意思是说，烛薪烧尽，火种却一直流传下去。后来人们以"薪尽火传"这个成语典故，比喻学问和技艺代代相传。应该说，李启贵是有远见的，他正值盛年的时候就已经把技艺传承带徒弟的事纳入日程表了。现在活跃在各大饮食机构的名厨，有不少就出自他的门墙。这其中又有多少鲜为人知的故事呢？

In addition to the expensive Chinese Eight Treasures Tripod, the "Roast Duck with Ganoderma Lucidum" invented by Master Li Qigui was also an important dish in the Chinese Eight Treasures Tripod Banquet in Sunworld Dynasty Hotel. "There is Roast Duck in the restaurants I run. The dish Duck & Shark's Fin was Quanjude's famous dish that Master Cai Qihou personally taught me. As we all know, ganoderma lucidum has the effects of prolonging life, purifying blood and preserving health. When I was working in Osaka, I prepared sorghum-based liquors with ganoderma lucidum, which was very popular. In Sunworld Dynasty Hotel, I combined the ganoderma lucidum with the roast duck and prepared it into Ganoderma Lucidum Cake, Ganoderma Lucidum Sauce and Ganoderma Lucidum Duck. I pickled the duck with Ganoderma Lucidum and prepared Ganoderma Lucidum Sauce for it, which wasn't served elsewhere. What's more, I prepared ganoderma lucidum spore powder in the cake. All of these make it special."

The first dish in the "Chinese Eight Treasures Tripod Banquet" was Chinese Tripod, the second dish was Braised Longevity Sea Cucumbers with Scallion, and the third dish was Beijing Roast Duck with Ganoderma Lucidum, forming the "three pillars of the tripod". The main ingredients of Beijing Roast Duck with Ganoderma Lucidum included slaughtered lean meat stuffed duck, natural ganoderma lucidum, maltose water, scallion segments, sweet flour sauce, cucumber strips, mashed garlic, white sugar, etc. The process of Beijing Roast Duck with Ganoderma Lucidum mainly included such five parts as slaughtering and scalding, cut-opening, ganoderma lucidum soaking and processing, roasting and slicing duck. The difference between Beijing Roast Duck with Ganoderma Lucidum and other roast duck was that before roasting and after stuffing, ganoderma lucidum and ganoderma lucidum soup were poured, and then the roast duck was roasted after coloring. The furnace temperature was generally kept at 230 ℃ to 250 ℃, and the firepower should be adjusted at any time. After roasting, take out ganoderma lucidum for embellishment, and make Ganoderma Lucidum Sauce and Ganoderma Lucidum Cake with ganoderma lucidum powder, which were served together with scallion, cucumber strips, mashed garlic, white sugar and others as well as roast duck. This dish was characterized by crisp skin, tender meat, and tangy aroma, with functions of tonifying lung, benefiting stomach and tranquilizing mind.

In the heyday of Sunworld Dynasty Hotel, under the operation of Master Li Qigui, it was as "thriving" as the famous dish thriving for celebrating new yeas in Tang and Song Dynasties he studied and restored. However, how to better carry forward the Chinese food culture? Li Qigui thought more and did more than others.

Chuang Tzu said in an article entitled Nourishing the Lord of Life that, "Though the grease burns out of the torch, the fire passes on, and no one knows where it ends." When candles were not invented in ancient times, grease was used to wrap the firewood and light for lighting, and was called "candlelight". The sentence was later used as the idiom to analogically express the meaning that the torch of learning was passed on from teacher to student and from generation to generation. Li Qigui was far-sighted since he had already included the inheritance of skills and apprenticeship in his schedule in his prime. Many famous chefs now serving in major catering organizations once followed him. How many little-known stories were there?

1995年，北京丰泽园饭庄装修完成，重新开业。

In 1995, Beijing Fengzeyuan Restaurant was renovated and reopened.

1998年9月，李启贵被聘为北京市首届流水养鱼大奖赛总裁判长。

In September 1998, Li Qigui was appointed as the chief judge of the first Beijing Grand Competition of Running Water Culture.

1998年，中国烹饪协会副秘书长李正权先生为李启贵大师的中华八珍宝鼎提词。

In 1998, Mr. Li Zhengquan, Deputy Secretary General of the China Cuisine Association, presented an inscription for Li Qigui's Eight Delicacies Banquet.

1998年11月24日，李启贵大师率北京烹饪代表团到香港京华国际大酒店献艺表演，在香港引起轰动，23家新闻媒体争相报道。

On November 24, 1998, Li Qigui led a Beijing culinary delegation to Hong Kong to present a performance at the Metropark Hotel Kowloon, which caused a sensation in Hong Kong and was covered by 23 news media outlets.

2004年8月，李启贵大师和世烹联领导在天伦王朝接待了前来考察的西班牙大使。

In August 2004, Li Qigui and the leadership of the World Association of Chinese Cuisine received the Spanish ambassador at Sunworld Dynasty Hotel for a visit.

2008年3月，著名书画家徐悲鸿之子徐庆平为中华宝鼎专刊题词："中华第一鼎"。
In March 2008, Xu Qingping, son of the famous calligrapher and painter Xu Beihong, wrote an inscription for a special issue of the Chinese Eight Delicacies Banquet: "The No.1 Banquet of China".

中国烹饪协会第六届会长姜俊贤向李启贵大师颁发"改革开放40周年突出贡献奖"。
Jiang Junxian from China Cuisine Association presented Li Qigui with the Outstanding Contribution Award for the 40th Anniversary of Reform and Opening-up.

2009年10月，李启贵大师在扬州第十九届厨师艺术节成立名厨专业委员会，并荣获金牌奖。
In October 2009, Li Qigui established a professional committee of famous chefs at the 19th Chef Art Festival in Yangzhou and won the Gold Medal Award.

2010年2月4日，迎新春175个国家招待会在北京伯豪瑞庭五星级酒店举行。
On February 4, 2010, a reception for 175 countries to celebrate the Chinese New Year was held at the five-star Hotel Radegast Hotel Beijing Bohao.

2012年8月，李启贵大师在世界烹饪联合会举办的"中华厨艺高级研修班"上为韩国厨师颁奖。
In August 2012, Li Qigui presented an award to a Korean chef at the Advanced Chinese Culinary Arts Workshop organized by the World Association of Chinese Cuisine.

2001年，李启贵大师在天伦王朝酒店接待来访合作的法国、毛里求斯、墨西哥、乌拉圭四国大使和法国驻远东的主席古扎礼先生。

In 2001, Li Qigui received the ambassadors of France, Mauritius, Mexico and Uruguay and the Ambassador of France in the Far East, Mr. Gouzalet, at the Sunworld Dynasty Hotel.

李启贵大师在第七届中国烹饪世界大赛上评判菜肴。

Li Qigui judged dishes at the 7th World Championship of Chinese Cuisine.

中国十佳烹饪大师李启贵莅临满洲里国际饭店，欢迎仪式上合影。

Welcome ceremony of Li Qigui, one of China's top culinary masters.

2004年10月，中国烹饪协会领导和第五届全国大赛评委合影。
In October 2004, the leaders of China Cuisine Association and the judges of the 5th National Competition took a group photo.

2008年10月，世烹联第六届中国烹饪世界大赛评委合影。
In October 2008, a group photo of the judges of the 6th World Championship of Chinese Cuisine.

第五届全国烹饪技术比赛合影。
Group photo of the 5th National Culinary Arts Competition.

第七回

Chapter Seven

遍访名师 红白绝技都在手
感恩传递 无私课徒海胸襟

李启贵不仅有一双巧手，更有一双慧眼，堪称烹饪界的"伯乐"。谁是厨行的千里马，是"祖师爷赏饭"的那种人，李启贵上下打量几眼就能看出来。一旦他看上是块"好钢"，就会无私地传授自己的精湛技艺。英雄不问出处，才俊自有机缘。

李大师闲来无事喜欢逛潘家园，去挑选一些精巧别致的餐具和烹饪书籍。2000年的一天，李启贵在潘家园遛到中午1点半，就走进旁边的大鸭梨餐厅，准备在这儿吃午饭。他刚落座不久，餐厅有人就认出了他，说天伦王朝饭店的总厨师长来了。

过了一会儿，从后厨走出个年轻人，走到李启贵大师面前问："您是李大师吧？""啊，我姓李。"小伙子眼睛一亮，"我知道您，您很有名，我是这儿的普通炒菜厨师，叫胡国山，山东人。我想跟着您干。"李启贵表示机会成熟可以，于是两人相互留下了联系方式。隔了些日子，胡国山就去天伦王朝饭店找李大师了。当场一试，小伙子炒菜的路子还算周正，是个可塑之才，就收下了他。不久，北京功德福餐饮有限公司的永兴楼开业，胡国山就被李启贵大师推荐去做了厨师长。收徒就要传道授业解惑，李启贵从不做挂名的师父，也不收寄名的徒弟。于是，李大师就开始教胡国山雕刻厨艺。厨艺雕刻分几个层面，既有"果蔬雕""黄油雕"，也有"冰雕"和"面塑"等多种形式。

李启贵不但教能耐，还送趁手的家伙。他有三套冰雕的工具，一套自用，一套给了功德福的厨师长张国，一套赠给了胡国山。这"冰雕三件套"，一件是用一寸的刨子刃儿焊在一根钢筋上，横着焊成一个丁字形；另外两件，是两把镩冰雕刻用的刀子，一把一寸五的，一把两寸五的。

胡国山如获至宝，按照师父的要求真下了功夫了，光练瓜雕就雕了几百个西瓜。所谓学会文武艺，货卖与识家。又过了一段时间，胡国山跟师父李启贵说："师父，我也想出去闯荡一下，南下无锡，检验一下自己的手艺。"结果到了无锡没干多长时间，一个多月就回来了，说："南方不认咱北方菜，这次

Visiting Many Famous Masters and Mastering Dish-cooking & Noodle-making Skills
Passing on Unique Skills with Gratitude & Delivering Unselfish Lessons

Li Qigui not only boasted extraordinary skills, but also developed a sharp eye for discovering able people, so he could be called a "talent scout" in the culinary field. Li Qigui could find potential cooks who were destined to be excellent chefs at a glance. Once he found the "good candidates", he would selflessly teach his exquisite cooking skills. Regardless of origin, talents were prepared to enjoy opportunities.

Master Li liked to visit Panjiayuan market in spare time, choosing some exquisite and unique tableware and cooking books. One day in 2000, Li Qigui strolled in Panjiayuan market until 1:30 p.m. and walked into the nearby Dayali Restaurant to have lunch there. Shortly after he took his seat, someone in the restaurant recognized him and said that the chief chef of Sunworld Dynasty Hotel had arrived.

After a while, a young man walked out of the back kitchen, walked up to Master Li Qigui and asked, "Are you Master Li?" He replied, "Yes, my surname is Li." The young man's eyes lit up, "I know you, you are very famous, I am an ordinary cooking chef here, named Hu Guoshan, and I am from Shandong. I want to learn cooking from you." Li Qigui said it's possible if having chance, so they left contact information with each other. A few days later, Hu Guoshan went to Sunworld Dynasty Hotel to visit Master Li. On the spot, the young man was knowledgeable about cooking and was a potential chef, so Master Li accepted him as an apprentice. Soon, Yongxing Restaurant of Beijing Gongdefu Catering Co., Ltd started operation, and Hu Guoshan was recommended by Master Li Qigui to be the chief chef. To accept apprentices, the master must transmit wisdom, impart knowledge, and resolve doubts. Li Qigui never became a nominal master, nor did he accept nominal apprentices. Therefore, Master Li began to teach Hu Guoshan carving cooking. Cooking carving was divided into several levels, including "fruit and vegetable carving", "butter carving", as well as "ice carving" and "dough carving" and other forms.

Li Qigui not only taught skills, but also sent them good presents. He had three sets of ice carving tools, one for his own use, one sent to Zhang Guo, the chief chef of Gongdefu and one to Hu Guoshan. These "three sets of ice carving tools", one was welded to a steel bar with an inch plane blade and welded horizontally into a T-shape; the other two were two knives for cutting ice sculpture, one was 1.5 inch and the other was 2.5 inch.

Hu Guoshan treated it as a precious jewel in his hands, so he really worked hard according to Master's request. He carved hundreds of watermelons just by practicing melon carving. With professional skills, one would be appreciated by connoisseurs. After another period of time, Hu Guoshan said to Master Li Qigui, "Master, I also want to go out for a trial, and go south to Wuxi to test my cooking skills." It turned out that the southerners didn't recognize northern cuisine after he worked over a month. He came back and asked his master to give some advice. Li Qigui transferred him to the Cantonese kitchen of Sunworld Dynasty Hotel and created a good working environment and opportunities for him. Following the ice sculpture cold storage of Tokyo Prince Hotel in Japan,

闯荡不成功，您看我干点什么吧。"李启贵就把他调到天伦王朝饭店的粤菜厨房，并且给他创造了一个很好的工作环境和机会。照着日本王子饭店的冰雕冷库，李启贵花了好几万在天伦王朝也造了一座，连冰雕的起降车、一应工具都置办齐了。胡国山到了这儿，简直如鱼得水，一头钻进冷库研究冰雕技艺，经过多年的努力，后来终于成为一代冰雕大师。

后来胡国山自己以《伯乐相马》为题，写下了一段发自肺腑的感言，他先引用了一段韩愈的《马说》，然后套用其句式写道："世有名厨大师，然后才有厨师。厨师很多，但是大师却很少。即使是很有才华的厨师，也只能在小餐馆里劳作，挥汗如雨，卖命于锅碗瓢盆之间，拼死于火烤火燎之中，不以名厨称也。但是遇到一位好师父，命运却改变了。学厨艺，学厨德，学做事，学做人。世界各国、全国各地的有志之士都争相赶来投奔，却又有很少的一部分有缘的人才收入门下。机遇难求。师父真可谓'德艺双馨，桃李满天下'。他带领我们百名徒弟，振兴中华餐饮，发扬师父精神，继承师父厨艺，横扫江湖，名震四方，风靡世界。徒弟胡国山。2013年6月27日。"

说起这些，李启贵说，我自己又何尝不是呢？我的这些厨艺也不是天生就会的，是当年我的师父王义均大师和其他老师傅们的无私教诲，才使我一步一步走到今天。"所以我常常扪心自问，什么叫感恩？感恩不是我逢年过节提着东西去看望教过我技艺的老师们那么简单。而是我能不能像当年师父教我那样，无私地、热情地往下传授给我的徒弟们。传技艺、传做人，这样才叫感恩。"

李启贵说，他最早的开蒙老师艾长荣教过他做"炸虾球"，这是鲁菜中的一道名菜。"炸虾球"是把虾米制成泥子，里面加上蛋清、团粉、盐味、葱姜，很香很好。虽然很香嫩，但是实心的。1993年李启贵参加第三届全国烹饪大赛，为了厨艺比别人多一手，他的师父王义均又教了他一道"空心虾球"。

当时李启贵住在右安门内一座六层楼的一居室里。房子小，厨房更小，可是他的师父王义均、李正权两位老先生，多少次爬上六楼，钻进他的小厨房里，手把手地教他研制"空心虾球"。多年后，年届七旬的李启贵大师提起此事，用八个字概括："师恩浩荡，饮水思源"。

王义均大师当时给他讲了一个道理：你知道一个菜，听人说了一个菜，你学会了一个菜，这是三个过程。你知道了一个菜，听人说10遍，不如眼见一

Li Qigui spent RMB tens of thousands Yuan to build one storage, and even bought all the tools for the ice sculpture including landing vehicle. When Hu Guoshan arrived there, he felt just like a fish in water. He plunged into the cold storage to study the skills of ice sculpture. After years of hard work, he finally became an ice sculpture master of the generation.

Later, Hu Guoshan himself wrote a heartfelt speech under the topic of *Bole Judging Horses*. He first quoted a passage pattern from Han Yu's On Horses with his words, "Only after the cooking master come into the world are there chefs able to cook excellent dishes. Such chefs are common, but a cooking master is rare. So even talented chefs, if slaved by small restaurants, will work themselves to the bone and perish in their kitchens without being known as good chefs. But the fate will change once luckily meeting a good master so as to learn cooking skills and ethics, learn to handle affairs, and learn to grow up. People with lofty ideals from all over the world rushed to learn cooking skills from Master Li, but only a small number of predestined talents are accepted. It's really rare opportunities. Master Li is excellent in both performing skills and moral integrity and has students all over the country. He is leading our 100 apprentices to revitalize Chinese catering. We should carry forward Master's spirit and inherit Master's cooking skills, making them well-know all over the world. Apprentice Hu Guoshan. June 27, 2013."

Speaking of this, Li Qigui said it was the same to him. He was not born with cooking skills but got them from the selfless teachings of his master Wang Yijun, and other old masters who made him achieve this day step by step. "So I often ask myself, what is gratitude? Gratitude is not as simple as carrying things to visit teachers who have taught me skills at festivals. But can I impart to my apprentices selflessly and enthusiastically as master taught me in those days. Transmitting cooking skills and behaving in a proper way is gratitude."

Li Qigui said that his earliest instructor Ai Changrong taught him to cook Fried Shrimp Balls, which was a famous dish in Shandong cuisine. The Fried Shrimp Balls was made as the following: make shrimp into paste, add egg white, ball powder, salt, scallion and ginger. It tasted fragrant and good. Although it was fragrant and tender, it was solid. In 1993, Li Qigui took part in the 3rd National Cooking Competition. In order to cook better than others, his master Wang Yijun taught him another Hollow Shrimp Balls.

At that time, Li Qigui lived in a one-bedroom house in a six-story building in Youanmen Inner Street. The house was small and the kitchen was even smaller, but his masters, Wang Yijun and Li Zhengquan, climbed to the sixth floor, got into his kitchenette and taught him to make Hollow Shrimp Balls hand in hand. Many years later, Master Li Qigui, in his seventies, mentioned the matter and summed it up as follows, "Mighty kindness of instructors wouldn't be forgotten".

Master Wang Yijun told him a truth at that time: You knew a dish, heard about a dish, and learnt a dish. These were three processes. It was better to see a dish once than heard about it 10 times. It was better to cook it once than seeing 10 times. It was better to cook it 10 times than to cook it once. The feeling that you were familiar with was far from the feeling that you cooked it carefully. It would be two processes when you cooked it carefully and skillfully. Therefore, if you want to reach the level where you were skillful, you must practice every day.

In fact, such precious words could be applicable everywhere without exception. During that period, Li Qigui learned to make "Hollow Lobster Balls" at the rhythm of making one lobster every other day and made countless lobsters. More than a month before the competition, Li Qigui invited Master Wang at home and cooked the Hollow Lobster Balls on the spot, and then his master

遍；眼见10遍，不如动手干一遍；动手干一遍，不如动手干10遍。你干熟了的感觉，跟你干细了的感觉相差很远。你干细了跟你干精了到你伸手就有的感觉，又经过了两个过程。所以你要想达到伸手就有的程度，你必须要天天练。

其实这样的金玉良言，可以放之四海而皆准，学什么都概莫能外。那段时间，李启贵学做"空心龙虾球"是隔一天做一只的节奏，做了无数只龙虾。离开赛还有一个多月，李启贵把师父请到家去，当场就做，吃了就评。隔一天备料，第二天接着做。最后等到他在石家庄上赛场的时候，轻轻松松就把这个菜完成了，盛菜的那个镀金盘子还是李启贵用工资奖金买的，没有动用公家的任何东西。这次大赛，李启贵凭借"空心龙虾球"和"茉莉凤蓉竹荪汤"两道菜，荣获了金奖第一名，同时获得"全国优秀厨师"的称号。

荀子有云："青，取之于蓝，而青于蓝；冰，水为之，而寒于水。"李启贵精湛的厨艺来自于名师的悉心指导，也来自于用心地吸收消化和创新。我们不妨详举一例。比如王义均大师曾经教过李启贵一道"扒蝴蝶海参"的名菜，多少年后又以一种全新的面貌呈现出来。

"蝴蝶海参"的菜名记载始见于明代。原来指海参直接抹刀片即为蝴蝶片，抹泥子后以蝶为形似，不但时常脱落，而且远未逼真。王义均大师别出心裁，显著提升了此菜的造型美。他使用的原料包括水发海参、鸡脯肉、鸡蛋白、金华火腿、蛋糕、香菜梗、水发银耳、油菜心、高汤、精盐、味精、料酒、姜汁、淀粉、黑芝麻、青红椒米、木耳米、蛋糕米、葱姜油。

制作过程是将水发海参洗净，做成12个蝴蝶形的片，用热水氽透。将鸡脯肉打成泥状，装入塑料袋中备用。选香菜细梗洗净，切成3厘米长的段，火腿、蛋糕切成1厘米的丝，木耳、青椒切成1厘米的细丝。蛋糕丝需要12根，全体细丝各需24根，将保鲜纸铺在干净配菜盘上。将盛放鸡泥的袋子剪去一角，用手将泥挤在保鲜纸上，挤成12个略长的蝴蝶肚形，约3厘米。把做成蝴蝶形的海参片分别放在泥子上，上边再挤一层泥子，将蝴蝶海参两头粘好（使其呈蝴蝶状），前边分别插上两根香菜梗（作为须子），在上边粘上两颗黑芝麻（表示眼睛），把蛋糕丝、火腿丝、青椒丝粘在蝴蝶肚子上，上屉蒸3分钟取出。将银耳切成块，拌好鸡泥子，用开水氽熟。将油菜心洗净煸出并入味，根朝里码放在盘子内，呈放射状，中间放上银耳。蒸好的蝴蝶海参码在盘子周围的油菜叶上。炒锅上火，放入高汤，勾成水晶汁（扒汁），淋在整个菜品上。炒锅刷

commented on it after having it. He cooked one day after preparing ingredients. Finally, when he took part in the competition in Shijiazhuang, he finished the dish easily. The gold-plated plate serving the dish was bought by Li Qigui with salary bonus instead of using company funds. In that competition, Li Qigui won the first gold medal and the title of "National Excellent Chef" with "Hollow Lobster Balls" and "Dictyophora Soup with Jasmine and Amaranth".

Hsun Tzu said, "The dye extracted from the indigo is bluer than the plant, so is the ice colder than the water." Li Qigui's exquisite cooking skills developed from the careful guidance of famous teachers, as well as careful absorption, digestion and innovation. We might as well give a detailed example. For example, Master Wang Yijun once taught Li Qigui a famous dish of "Braised Butterfly-shaped Sea Cucumbers". Many years later, it took on a completely new look.

The name of "Braised Butterfly-shaped Sea Cucumbers" was recorded in the Ming Dynasty. It originally referred to sea cucumbers directly cut into butterfly oblique slices. After being wrapped with paste, they look like butterflies, which not only often fell off, but also were far from realistic. Master Wang Yijun's ingenuity had significantly improved the beauty of this dish. The main ingredients he used included water-fat sea cucumbers, chicken breast, egg white, Jinhua ham, cake, coriander stalk, water-fat tremella, rape, soup-stock, refined salt, monosodium glutamate, cooking wine, ginger juice, starch, black sesame, green and red pepper grains, tremella grains, cake grains, and fried scallion-ginger oil.

The process of Braised Butterfly-shaped Sea Cucumbers was as follows: wash the water-fat sea cucumbers with water, cut 12 butterfly-shaped slices, and boil them thoroughly in hot water. Mince the chicken breast and put it into plastic bags for later use. Choose coriander stalks, wash and cut them into 3cm long segments, cut ham and cake into 1cm shreds, and cut agaric and green pepper into 1cm shreds. 12 pieces of cake shreds and 24 pieces of shreds for all others are needed. Spread fresh-keeping paper on a clean side dish. Cut off one corner of the bag containing minced chicken breast, squeeze it on the fresh-keeping paper by hands as 12 slightly longer butterfly belly, about 3cm. Place the butterfly-shaped sea cucumber slices on the minced chicken breast respectively, squeeze another layer of minced chicken breast on the top, stick the two ends of the butterfly-shaped sea cucumbers (make it butterfly-shaped), insert two coriander stalks (as beards) on the front, stick two black sesame seeds (for eyes) on the top, stick cake shreds, ham shreds and green pepper shreds on the butterfly belly, steam in the drawer for 3 minutes and take out. Cut tremella into pieces and mix them with minced chicken breast and boil them in boiling water. Wash the rape and stir-fry it until it was flavored. Stack it on a plate with the root inward in a radial shape with tremella in the middle. Stack steamed butterflies and sea cucumbers on rape leaves around the plate. When the frying pan was heated, add the soup-stock, cook it into crystal juice (braising juice), and pour it on the whole dish. Brush the frying pan clean, add fried scallion-ginger oil, red pepper grains, green pepper grains, tremella grains and cake grains, and sprinkle them evenly on the tremella in the middle. This dish was beautiful in shape and tasted light and palatable.

Master Li Qigui's practice was different from that of his master. The name of the dish was changed to "Green Butterflies with Minced Shellfish in Clear Soup". The main ingredients were natural dictyophora, fresh shellfish, chicken tenderloin, shark's fin needle, green vegetables, cucumber, carrot, water-fat mushroom, egg white, clear soup, refined salt, cooking wine, ginger juice and water starch. The preparation method was to remove the skin of fresh shellfish and chicken tenderloin, add refined salt, cooking wine, ginger juice, egg white liquid and water starch, beat them into paste, add green

净，放入葱姜油和红椒米、青椒米、木耳米、蛋糕米煸热，均匀地撒在中间银耳上即成。这道菜造型美观，清淡适口。

　　李启贵大师的做法与他的师父比起来又有所不同。菜名改为"清汤贝蓉翡翠蝴蝶"。原料选用上好的天然竹荪、鲜贝、鸡里脊、鱼翅针、青菜、黄瓜、胡萝卜、水发香菇、鸡蛋清、清汤、精盐、料酒、姜汁、水淀粉。制法是将鲜贝、鸡里脊去筋皮，加入精盐、料酒、姜汁、蛋清液、水淀粉打成蓉，取一半加上绿色菜汁搅拌均匀。竹荪发好制成蝴蝶形，抹上绿色的贝蓉，用鱼翅针作须，胡萝卜、香菇、黄瓜等切丝作背，上屉蒸熟。另一半贝蓉挤成圆形汆熟。清汤加调味料，烧开后去浮沫，盛入汤窝中。放入蒸好的"蝴蝶"和贝蓉圆即成。这道菜的特点是白绿相间，形象美观，汤鲜味爽。从菜形上来说，李启贵大师制作的"清汤贝蓉翡翠蝴蝶"同样赏心悦目。

　　还有一道菜"葱烧海参"也是如此，李启贵在师父王义均传授技艺的基础上，守正创新，衍生出了一道"葱烧万寿参"。这道菜的主料是水发辽参。配料是葱段、油菜心、枸杞、南瓜。调料包括清汤、葱油、精盐、白糖、料酒、酱油。先将南瓜去皮刻成"寿"字备用。再将油菜心根改成十字花刀。枸杞发好备用。然后把海参洗净放入锅中，加入清汤、料酒、精盐、酱油、糖、葱油煨透入味。另起锅炒糖色，将海参放入锅内烹料酒放清汤，加料酒、糖、精盐、葱油，小火慢煨，同时将葱段炸好、上屉蒸好。菜心煨好码在盘中，下面放蒸好的葱段，南瓜寿字蒸好码在盘子周围，浇上汁。海参已煨透入味，滚芡点葱油放在盘子中间，放青蒜段即可。这道菜的特点是海参葱香味浓，柔软润滑，造型美观。给老人做寿宴，极富喜庆色彩。

　　蔡启厚先生教的几道菜包括"猴头燱大虾""一品芙蓉""金鱼系明珠"。其中"猴头燱大虾"是夺得1986年卢森堡国际奥林匹克烹饪大赛金奖的获奖作品。"一品芙蓉"非常精美，如诗如画。

　　"金鱼系明珠"则富有童趣。这道菜的主料是对虾，配料是鸽子蛋、鸡里脊、青笋、水发香菇、胡萝卜、黄瓜、蛋清。调料有精盐、料酒、葱姜油、胡椒粉、水团粉、白糖、清汤。大虾去皮留尾，一分为二，后边带尾打花刀入味，前边去头斩成蓉，与鸡蓉一起调味，加入蛋清、水团粉，用尺板抹成鱼形。用胡萝卜和青笋、黄瓜做嘴、眼睛和鳞片，鸽子蛋煮熟去皮入味，码在盘子中间，全鱼蒸熟摆在周围，浇白汁即可。此菜造型栩栩如生，大虾肉鲜味美，赏心悦目。

vegetable juice to half, and stir evenly. The water-fat dictyophora was made into butterfly shape and smeared with green shell paste, shark's fin needles were used as whisker, and carrot, mushroom, cucumber, etc. were shredded as back, and steamed in a drawer. The other half was squeezed into a circle and boiled. Add seasonings to the clear soup, boil, remove floating foam, and put it into the soup nest. Add steamed "Butterfly" and minced shellfish. This dish was characterized by white and green with beautiful image and delicious soup. In terms of dish shape, Master Li Qigui's "Green Butterflies with Minced Shellfish in Clear Soup" was also pleasing to the eye.

The same is as another dish "Braised Sea Cucumbers with Scallion". On the basis of the skills taught by Master Wang Yijun, Li Qigui kept the right track and innovated, developing Braised Longevity Sea Cucumbers with Scallion. The main ingredient of this dish was water-fat sea cucumbers from Liaoning. Ingredients were scallion segments, rape, Chinese wolfberry and pumpkin. Seasonings included clear soup, scallion oil, refined salt, white sugar, cooking wine and soy sauce. Peel the pumpkin and carve it into the word "longevity" for later use. Cut the rape root into cruciform pattern. Soap Chinese wolfberry in water for later use. Then wash sea cucumbers and put them into a pan. Add clear soup, cooking wine, refined salt, soy sauce, sugar and scallion oil to simmer until it's tasty. Stir-fry sugar in another pan. Put sea cucumbers into the pan, and cook with cooking wine and clear soup, and then add cooking wine, sugar, refined salt and scallion oil. Simmer slowly over a very low flame. At the same time, fry scallion segments and steam them in a drawer. Place simmered rape over steamed scallion segments in the plate, and place steamed pumpkin with longevity character in the plate with juice around. The sea cucumbers were simmered until tasted delicious. Thicken them with starch, add scallion oil and put them in the middle of the plate and then add leeks and garlic segments. This dish was characterized by thick fragrance, softness and smoother of sea cucumbers and scallions, and beautiful in shape. It is very festive to give the aged a birthday party.

Several dishes taught by Mr. Cai Qihou included "Simmered Prawns with Hericium Erinaceus", "Top-graded Lotus" and "Goldfish with Pearl". Among them, "Simmered Prawns with Hericium Erinaceus" enabled Li Qigui to win the gold medal in the IKA in 1986 in Luxembourg. "Top-graded Lotus" was very exquisite and picturesque.

"Goldfish with Pearl" was full of children taste. The main ingredient of this dish was prawns, and ingredients included pigeon eggs, chicken tenderloin, green bamboo shoots, water-fat mushrooms, carrots, cucumbers and egg white. Seasonings included refined salt, cooking wine, fried scallion-ginger oil, pepper, water powder, white sugar and clear soup. The prawns were peeled with tails left, and divided into two parts. The back part with the tail was cut in a cruciform pattern for tastiness. The front part was minced into paste after the head was removed, season with minced chicken, add egg white and water powder, and make it into fish shape with a ruler. Use carrots, bamboo shoots and cucumbers as mouths, eyes and scales. Boil, peel and savor pigeon eggs. Place them in the middle of the plate. Steam the whole fish and pour white juice around them. This dish was lifelike in shape and pleasing to the eye, with delicious prawn meat.

The famous dishes that Li Qigui learned from the old master Ai Changrong and then made innovation included "Fresh Seafood with Sauce" and "Braised Eight Seafood in All". The dish "Fresh Seafood with Sauce" tasted very wonderful. The main ingredients included a live plectropomus leopardus, water-fat sea cucumbers, fresh squid, fresh shrimps, scallion, ginger, leeks, cooking wine, refined salt, oyster sauce, clear soup, water starch, soy sauce, edible oil, etc. The method was as follows: kill the live plectropomus leopardus, cut it into uniform pieces obliquely, thicken them with

李启贵从艾长荣老师傅手上学来又推陈出新的名菜有"浇汁海上鲜""全爆海八珍"等。"浇汁海上鲜"的菜品十分精彩，原料包括鲜活东星斑一尾、水发海参、鲜活鱿鱼、鲜活虾、葱、姜、青蒜、料酒、精盐、蚝油、清汤、水淀粉、酱油、食用油等。做法是将鲜活的东星斑宰杀好，改成均匀的翻刀片，挂上水淀粉，用热油炸好放入盘中，海参改成抹刀片，鱿鱼切鱼鳃花刀，葱切丝，姜切末，放入碗中加清汤、调味品对成汁。海参、虾、鱿鱼滑好炒汁，下入配料，浇在炸好的鱼上即可。此菜外焦里嫩，海鲜浓郁，口感丰富。

勤奋多思的李启贵站在烹饪巨人们的肩上发展，自然与众不同。他跟马德明师傅学的"通天鱼翅""清汤官燕""芙蓉鸡片""炒龙凤丝""糟熘三白""油爆双脆"等，至今脍炙人口，经久流传。"炒龙凤丝"的原料是鳜鱼肉、鸡脯肉、葱姜油、精盐、料酒、蛋清、水淀粉、清汤、胡椒粉、食用油等。制法是将鳜鱼肉、鸡脯肉匀切成丝（鱼丝略粗，鸡丝略细），分别用蛋清、水淀粉拌匀浆好，然后用温油滑好，控油备用，炒锅上火，倒入滑好的鸡丝、鱼丝，放入精盐、料酒、清汤、胡椒粉，颠翻均匀，加入葱姜油即成。这道菜以鱼为龙，以鸡为凤，故称"炒龙凤丝"，色彩白净，入口滑嫩，清鲜味爽。

说起面点圣手周子杰，李启贵的感激之情溢于言表。他说，周老爷子不但厨艺高超、手法独特，而且人品上乘。当年都是手把手地教他能耐，今天做荞麦棱儿，明天做一窝丝，后天教做韭菜扁儿，大后天教"雪花龙须面"……"老爷子抻面没有面头儿，这是盖世无双的。甭管多少面做完，就剩二两面头儿，最后一揉一擀下到锅里，煮出半碗自己吃了。"李启贵说起这个来，眼神里充满了折服和敬佩。

要留住中华优秀的烹饪文化传统，就要"薪火相传"，李启贵从周子杰那儿学到的抻"雪花龙须面"的绝技，现在已经传给了两位高徒。一位叫周占强，一位叫陈彩霞。如今他们都是面点方面一等一的高手。陈彩霞在2008年北京奥运会上获得了烹饪特别金奖。周占强则在天伦王朝饭店举行的38家下辖酒店烹饪大赛中获得了面点组的金奖，经常上电视表演面点技艺。他和陈彩霞现在都是伯豪瑞庭五星级酒店的名厨。

李启贵从以王义均为主的诸多厨艺大师们的身上，学到了5000年中华烹饪文化的精髓，同时又能结合社会的进步、时代的变迁，不断推陈出新，守正创新，在国内外的烹饪舞台上屡夺魁元，获奖无数，这其中有什么奥妙吗？若干年后，他能把这些"一招鲜"和"必杀技"，毫无保留地传授给他的弟子们吗？新一代京鲁菜的传人们能否继续李启贵开创的辉煌呢？且听下回分解。

water starch, fry them in hot oil and put them into a plate, cut sea cucumbers into slices obliquely, cut squid into fish gill shape, shred scallion, mince ginger, put them into a bowl and add clear soup and condiments to make sauce. Fry sea cucumbers, shrimps and squid, add the ingredients and pour the sauce on the fried fish. This dish was tender with a crispy crust, with strong seafood flavor and rich taste.

Diligent and thoughtful Li Qigui developed on the shoulders of cooking giants. Naturally, he was different from others. He learned such dishes from Master Ma Deming as "Supreme Shark's Fins", "Fine White Cubilose in Clear Soup", "Chicken Slices with Lotus", "Fried Dragon and Phoenix Shreds", "Sauteed Chicken", "Fish and Bamboo Shoots with Rice Wine Sauce" and "Fried Double Crispy Ingredients with Oil", which are well-known and have been circulated for a long time. The main ingredients of "Fried Dragon and Phoenix Shreds" were mandarin fish meat, chicken breast, fried scallion-ginger oil, refined salt, cooking wine, egg white, water starch, clear soup, pepper, edible oil, etc. The preparation method was to evenly cut mandarin fish meat and chicken breast into shreds (fish shreds were slightly thicker and chicken shreds were slightly finer), mix egg white and water starch respectively, then smooth them with warm oil, and drain oil for later use; heat stir-fry pan, pour the smoothed chicken shreds and fish shreds, add refined salt, cooking wine, clear soup and pepper, turn over evenly, and add fried scallion-ginger oil. This dish took fish as dragon and chicken as phoenix, so it was called "Fried Dragon and Phoenix Shreds". It was white in color, smooth and tender in taste, and clear and delicate in flavor.

Speaking of pastry master Zhou Zijie, Li Qigui's gratitude was beyond words. He said that Zhou not only had excellent and unique cooking skills, but also had outstanding personality. In those days, he was taught hand in hand one by one, such as Prismatic Buckwheat Pieces, Coil Cake, Leek Roll, and "Snowflake Longxu Noodles" ... "The old master pulls noodles without ends, which is unparalleled. No matter how many noodles are pulled, there are only two ends left. Finally, he kneads them and boils in the pot for himself." Speaking of this, Li Qigui's eyes were filled with submission and admiration.

In order to retain Chinese excellent cooking culture tradition, it is necessary to "pass on the skills from generation to generation". Li Qigui's unique skill of pulling "Snowflake Longxu Noodles" learned from Zhou Zijie has now been passed on to two senior apprentices. One was Zhou Zhanqiang and the other was Chen Caixia. Now they are all first-class pastry experts. Chen Caixia won the special gold medal in cooking at the 2008 Olympics in Beijing. Zhou Zhanqiang won the gold medal in the noodle-making group in the cooking competition of 38 hotels under the management of Sunworld Dynasty Hotel, and often performed noodle skills on TV. He and Chen Caixia are now famous chefs in the five-star hotel of Radegast Hotel.

Li Qigui had learned the essence of Chinese culinary culture for 5,000 years from many culinary masters, mainly Wang Yijun. At the same time, he could combine the progress of society and the changes of times, constantly innovating while keeping the right path. He won numerous first prizes and awards on the culinary stage at home and abroad. Is there any secret behind this? A few years later, could he pass on these unique and essential skills to his apprentices without reservation? Could the new generation of Beijing-Shandong cuisine cook continue the glory created by Li Qigui? Please find answers in the next chapter.

1987年，李启贵大师正在丰台六分部讲授烹饪理论课程。
In 1987, Li Qigui was teaching a culinary theory course at the sixth division in Fengtai.

1987年，李启贵大师在烹饪班上示范菜肴。
In 1987, Li Qigui taught full-time.In 1987, Li Qigui was demonstrating the cooking of dishes in a cooking class.

1987年，李启贵大师专职教学。
In 1987, Li Qigui taught full-time.

1988年，李启贵大师在宣武烹协传授烹饪技艺。
In 1988, Li Qigui taught cooking skills at Xuanwu Culinary Association.

李启贵大师做菜。
Li Qigui was cooking.

1987年，李启贵大师正在亲自操作菜肴。
In 1987, Li Qigui was preparing dishes himself.

李启贵大师做菜。
Li Qigui was cooking.

中国烹饪名家 —— 李启贵：京菜

第八回

Chapter Eight

良乡传艺 授徒先从磨刀起
克绍箕裘 弟子烹坛屡夺魁

———

李启贵大师收徒弟的标准，在京城厨行可是以"严苛"闻名的，不能吃苦的，脑子不够使、祖师爷不赏饭的，偷奸耍滑人品不济的，一概不要。

李启贵收的第一个徒弟叫贾河武。那是在20世纪80年代初，已经小有名气的李启贵跟着宣武区烹饪协会到房山县的良乡去讲烹饪课。宣武区烹饪协会成立得比较早，协会的负责人叫张越，办事员叫李光贤。协会经常组织宣武区各餐饮单位的名厨到郊区去授课传艺，光李启贵去过的区县就有房山、门头沟、昌平、丰台、石景山、大兴、怀柔等。听课的学员都是各区餐饮企业有培养前途的青年才俊。贾河武就是其中的一位，当时他只有十七八岁。

说起来贾河武学厨艺是有家学渊源的。他的父亲新中国成立前就在良乡十字大街路口干饭馆，叫良乡饭馆，后来公私合营了改叫良乡一部。也就是现在北京公德福餐饮有限公司的前身。如今的贾河武克绍箕裘，努力经营着这个房山最大的餐饮集团。

贾河武所在的学员班有四十多人，他算这个班里学习成绩比较好的一个。学员班的班主任由良乡饮食公司的教育科科长米子江兼任。米子江身高将近一米八九，人称"大米"。当时先办的是初级班，就是从学厨师入门开始，切葱、切蒜、砸毛姜。姜要切好了，得叫"姜米"。切葱有多少种刀法，蒜怎么切，都有门道儿。贾河武是学员中比较用功的。

回忆起当时的教学情景，李启贵大师记忆犹新。他说："切姜米得从头教起，生姜去皮，然后切片，顶刀切成丝，再顶刀切成米。我和贾河武认识就是从切姜米开始的。他切的姜米都带毛儿。什么原因呢？因为姜都有筋丝，刀不快就切出来带毛儿。快刀切出来的方方正正，跟米粒大小差不多。贾河武切出来的是一堆毛茸茸的小粒儿。我问他知道是怎么回事吗？他说我还真不知道。"

李启贵耐心地告诉他："我先教你磨刀吧。先用粗石磨，把你的刀刃儿给开出来，你的刀太钝了。刃儿开出来，再用细石磨两遍，磨出锋利来。现在老说砥砺奋进，砥就是粗磨刀石，砺就是细磨刀石。"李启贵边教边演示，磨刀

Passing on Culinary Skills from Liangxiang and Starting from Sharpening Knives
Carrying Forward the Cooking Skills and Empowering Apprentices to Win Medals

Master Li Qigui followed "high demanding" standards for accepting apprentices, which had been well-known in the Beijing cooking industry. Those who could not bear hardships, behave flexibly and have talent, but had poor moral qualities such as stealing, cheating, or playing know-hows would not be admitted.

Li Qigui's first apprentice is Jia Hewu. It was in the early 1980s that Li Qigui, who was already a minor celebrity, followed the Xuanwu District Cooking Association to Liangxiang, Fangshan County to give cooking lessons. Xuanwu District Cooking Association was established earlier. The head of the Association was Zhang Yue and the clerk was Li Guangxian. The Association often organized famous chefs from various catering units in Xuanwu District to give lectures and spread cooking skills in the suburbs. The districts and counties that Li Qigui had given lectures included Fangshan, Mentougou, Changping, Fengtai, Shijingshan, Daxing, Huairou, etc. Students attending the lectures were all young talents with promising training prospects from catering enterprises in all districts. Jia Hewu was one of them and then was only seventeen or eighteen years old.

As a matter of fact, Jia Hewu had a family origin in learning cooking skills. Before liberation, his father worked in a restaurant at the junction of Liangxiang Cross Street, which was called Liangxiang Restaurant and later changed to Liangxiang Department One after the public-private partnership was launched. It was the predecessor of Beijing Gongde Fu Catering Co., Ltd. Today Jia Hewu followed in his father's footsteps, trying his best to run the largest catering group in Fangshan.

The class Jia Hewu studied in had more than 40 students, and he was one of the students with better results in the class. Mi Zijiang, Director of Education Department of Liangxiang Catering Company, concurrently assumed the head teacher. Mi Zijiang was nearly 1.89 m tall and was nicknamed "Da Mi". At that time, a junior class was held first for students who started learning the basic knowledge about cooking such as cutting scallions, garlic and ginger. Ginger should be cut into minced ginger, being nicknamed "Jiang Mi". There were many kinds of methods for cutting scallions and cutting garlic was also a knack. Jia Hewu was one of the students who studies harder.

Recalling the teaching scene at that time, Master Li Qigui still remembered it vividly. He said, "To mince ginger, one must teach students from the beginning: peel, slice, shred and then mince ginger. Jia Hewu and I knew each other from mincing ginger. All the minced ginger he cut was hairy. What was the reason? Because ginger had fibres and strings, the minced ginger was hairy if the knife was not sharp. The minced ginger would be square while cutting with a sharp knife. Jia Hewu cut out a pile of hairy particles. I asked him if he knew what was going on. He said he really didn't know."

Li Qigui patiently told him, "Let me teach you how to sharpen your knife first. First, use a coarse stone to grind out a blade. Your knife is too blunt. Once the blade is ground out, then grind the blade twice with fine stone to sharpen it. Now to sharpen with coarse and fine stone is used to describe forging ahead in Chinese." Li Qigui demonstrated while teaching. The rule of sharpening the knife

的规矩是，丁字步往那儿一站，边上有水碗，往刀上撩水，磨刀讲究磨两头带当间儿。什么叫磨两头呢？就是刀往前一推磨刀尖儿，往后一拉磨刀根，然后再往前一推往后一拉，然后翻过来，倒把，往前一推，往后一拉。磨刀尖儿的时候，那个中间的刃儿也搭上了，磨刀根儿的时候，中间也搭上了，这就叫"磨两头带当间儿"。厨行有这么一句话，李启贵跟贾河武讲得很清楚，叫"墩子使边儿刀使尖儿"。如果你使的刀没有尖儿，在片东西或剁东西的时候，那个刀的前端跟后端不着墩子，切出来都是连刀断不开的。

自己的刀要保持刀的中间刃部分微有弧度最佳，厨师行管这叫什么呢？叫"关老爷的刀——圆刃了"。关羽的刀是青龙偃月刀，刀头是大月牙形的，那是战场上用的东西。厨师切东西的刀，最锋利的刃度跟墩子的平滑度不能吻合，就接触不上。刀的"肚子"粘到墩子了，刀根儿跟刀尖儿都够不着，所以造成了"连刀"现象。

过去常说，"战士的枪，厨师的汤"，其实厨师的刀又何尝不至关重要呢？贾河武弄懂了这些诀窍，就下了功夫苦练。李启贵从一开始教贾河武磨刀、切葱、切蒜、砸毛姜，到后来逐步教他切丝儿、丁儿、片儿、块儿、蓉、泥、粒，由浅入深。先学切片儿，切片儿拿刀熟了之后，开始教他"片儿薄、丝儿细、丁儿匀"，这个行内点术语叫"切肉片、剁肉丁、片肉丝、拉鸡丝儿"。拉鸡丝儿用的是"推拉刀"，这是刀工最讲究的。

20世纪80年代初，厨师常切的肉丝主要是鸡丝和肉丝两种。那么拉鸡丝和剁肉丝的粗细标准是什么呢？肉丝的要求是竹帘子的"帘子棍儿"，鸡丝的要求是"火柴棍儿"。剁肉丁，把一片肉先片到一公分厚，然后刀抡起来"砰"地一剁，这是一个方条儿。肉条顺过来再剁，每刀的距离都一样，就成了方丁儿。行话叫"骰子丁"。

片肉片是什么标准呢？比如说过去最盛行的滑熘里脊、木樨里脊、麻熘里脊。尽管这三个菜都是用里脊片，但是片的厚度不一样。因为木樨里脊近似于滑熘里脊，肉片可以略微厚一点儿。但是麻熘里脊不一样，它的标准叫"铜子儿片"，薄厚跟新中国成立前的铜钱或以前的二分钱钢镚差不多。刀法讲究极多，把坐板肉斜着45度角改成长条，顺刀改，顶刀切，切出来的叫"鹰嘴儿片"。做糖醋鱼、做浇汁海上鲜，切的是"翻刀片"。

鸡蓉怎么做呢？要把墩子刮干净，在墩子上敲打，然后用刀滗。滗就是把刀横过来，略呈45度。鸡蓉在滗的时候，在墩子跟刀当中出现了白筋，要把它

was to stand with a T-step, lift water on the knife from a water bowl on the side, and sharpen both ends with the middle part sharpened subconsciously. What did you mean sharpening both ends? That's, push the knife forward to sharpen the tip of the knife, pull back to sharpen the root of the knife, then push forward and pull back repeatedly, then turn over, turn back, push forward and pull back. When sharpening the tip of the knife, the blade in the middle was also put on, and when sharpening the root of the knife, the middle was also sharpened, which was called "sharpening both ends with the middle part sharpened subconsciously". There was such a saying in the cooking industry. Li Qigui and Jia Hewu made it very clear, "using the edge of piers and the tip of knife". If your knife had no sharp point, when cutting or chopping things, the front end and back end of the knife could not touch the pier, so you couldn't break the cuttings.

The best way to keep your own knife was to keep the middle edge of the knife slightly curved. What did the chefs call this? It was called "Lord Guan's Knife-Round Blade". Lord Guan's knife was a falchion with a large crescent head, which was used on battlefield. The sharpest edge of the chef's knife couldn't coincide with the smoothness of the pier, so they couldn't be contacted. If the "belly" of the knife was stuck to the pier, the root and tip of the knife could not reach it, thus causing the phenomenon of "connected cutting".

In the past, it was often said that "the cook's soup is as important as the soldier's gun". In fact, the cook's knife was also very important. Jia Hewu understood these know-hows and worked hard. Li Qigui taught Jia Hewu to sharpen knives, cut scallions and garlic and mince hairy ginger from the beginning. Later, he gradually taught him to cut shreds, dices, slices, pieces, minces, paste and grains from the basics to the sophisticated. Learn to slice at first and then learn to "slice and shred thinly, and dice evenly". These industry terms were called "cutting meat slices, chopping meat dices, shredding slices of meat, and pulling chicken shreds". Pulling chicken shreds was made by pushing and pulling chicken shreds, which was the most important in cutting skills.

In the early 1980s, chefs often cut meat shreds, mainly chicken and pork shreds. So what was the standard for thickness of chicken and pork shreds? The meat shreds should be as thick as the "curtain stick" of bamboo curtain, while the chicken shreds should be as thick as the "match stick". Dice meat, slice a piece of meat to 1cm thick first, and then chop it into a square strip, which would be chopped in a longitudinal direction, with the same distance between each chop, thus becoming cubes, called "dices" in jargon.

What was the standard for slicing meat? For example, tenderloin in white sauce, sweet clover tenderloin and tenderloin with pepper, which were the most popular tenderloin in the past. Although these three dishes were all made of tenderloin slices, the thickness of the slices was different. Because sweet clover tenderloin was similar to tenderloin in white sauce, the meat slices could be slightly thicker. However, tenderloin with pepper was different and should be as thick as a "copper coin" before liberation or a two-cent coin before. There were a lot of tips in cutting. Cut the meat of hindquarter into long strips at an angle of 45 degrees. Cut in a transverse direction first and then cut in a longitudinal direction, forming olecranon "V-shaped slices". As to Sweet and Sour Fish and Fresh Seafood with Sauce, the tenderloins should be cut obliquely.

How to make mashed chicken? Scrape the pier clean, tap the chicken on the pier, and then decant it with a knife. Decanting was to strain the chicken with the knife transversely at slightly 45 degrees. When decanting the minced chicken, white tendons which appeared in the chopping block and the

拣出来。肉粒要比肉丁小，肉末儿比肉粒又小一些，再细的就是蓉，最细的就是各种肉泥子。肉泥子就要过箩了。

李启贵对贾河武这样掰开了揉碎了传授，就是在延续中华饮食烹饪文化的传统。李大师深有所感地说："为什么说学抻面要三冬两夏，是适应面的软度跟硬度。学徒要三年零一节，也是这个意思，是一个学习逐步深入的过程。"

贾河武学完了这些基本刀功，李启贵就开始给他讲花刀。"比如切腰子就要用上麦穗花刀，刀的深浅度切好了之后，要用水焯，一遇热，腰子片都卷起来，像麦穗一样。腰子去完了筋皮，把腰心儿中间那点儿油去掉片开，左手往前一搓一攒，那腰臊就鼓出来了，一推一拉就片干净了。其实，好刀工就一下的活儿。这个刀工，叫作剞刀刀工。这是刀工里比较难的，因为斜刀片得要浅，立刀剁得要深，得掌握这个劲儿，不能给它剁断了，要剁到腰子厚度的五分之四，还有一点儿连着皮儿。片刀浅，立刀深，角度要合适，然后根据腰子的大小截刀分出块儿来，焯出来的腰花是立着的，一个个卷出来跟大麦穗似的。"

李启贵把这个技巧反复地讲给贾河武，并亲自示范。贾河武无数遍地反复练习，终于掌握了其中的诀窍。练好了"麦穗腰花"，再练"核桃腰花"，核桃腰花是俩直刀，深度也要掌握好了，切完了之后开水一焯，腰花得跟球似的那么圆乎。

腰花会切了，李大师再教切鱿鱼卷的"鱼鳞刀"。所谓"鱼鳞刀"就是横竖两溜儿坡刀，深度一样，横着截。还有一个"鱼鳃刀"，鱼鳃刀就是把腰花或鱿鱼正面儿切直刀，深度在鱿鱼厚度的五分之四。然后再横着片，第一刀横着一刀，不要片透，然后第二刀把它片透了，单刀不切到底，双刀切到底，开水里一焯，跟鱼鳃一模一样。

炝腰花是一道凉菜。把腰花切成"鱼鳃腰子"焯熟了，过凉水后码在一个碗里，上边搁上生菜，然后拿盘儿一扣，啪一翻，一大盘儿看着都是腰子，上面浇上三合油吃。还有"蛾眉腰丝"，是用刀横着片，立着切，属于另一种做法。

在李启贵看来，贾河武属于"祖师爷赏饭"的人才，所以教得格外细心。不但教主料的制作工艺，就连配料的用法也讲得细致入微。厨艺的一个基本要求是，丝儿配丝儿，片儿配片儿，丁儿配丁儿，块儿配块儿，而且配料不能大于主料，否则就喧宾夺主了。这盘菜端出来好不好看、协不协调、和不和谐，和主料、配料的搭配有很大关系。比如说使葱，就有主料、配料之分。葱用作

knife should be picked up. The minced meat was smaller than the diced meat and the ground meat was smaller than the minced meat. The finer meat was mashed chicken, and the finest was all kinds of meat puree. The meat puree should be go through the basket.

Li Qigui taught Jia Hewu in a detailed manner so as to carry forward the tradition of Chinese food and cooking culture. Master Li said with deep feeling, "Why does it take three winters and two summers to learn pulling noodles? It is to adapt to the softness and hardness of noodles. It's the same to the fact that apprentices should follow masters for three years and one season, which is a process of gradual and in-depth learning."

When Jia Hewu finished learning these basic cutting skills, Li Qigui began to teach him about cutting in a cruciform pattern. "For example, when cutting kidneys, you need to cut in a wheat spike pattern. With proper depth of cutting, the kidneys would be rolled up like wheat spikes when blanching it in hot water. After the kidney was stripped of its tendon and skin, the fat in the middle of the waist should be removed and then the kidney should be sliced. As soon as your left hand rubbed forward and clasped, the urine in the kidney would bulge out and the kidney would be sliced completely as soon as you pushed and pulled the knife. In fact, a good cutting chef would finish it at one stroke. This cutting skill was called carving skill, which was more difficult, because the oblique cutting should be shallow and the vertical cutting should be deep; it's hard to master this strength properly; otherwise the kidney would be cut off; it's better cut to the four-fifths of the thickness of the kidney and keep a little bit of it attached to the skin. The oblique cutting should be shallow and the vertical cutting should be deep with the appropriate angle. Then the kidney should be cut and divided into pieces according to its size. The blanched pieces of the kidney should be standing and rolled out like wheat spikes one by one."

Li Qigui repeatedly told Jia Hewu this skill and demonstrated it himself. Jia Hewu practiced countless times and finally mastered the know-how. After practicing "wheat spikes-shaped pieces of the kidney" and "walnut-shaped pieces of the kidney", the latter of which was made of two horizontal cuttings with proper depth and looked like round balls when the well-cut walnut-shaped pieces of the kidney were blanched in boiled water.

Then Master Li taught him the "fish scale-cutting skill" to cut squid rolls. The so-called "fish scale-cutting skill" was inclined cutting horizontally and vertically with the same depth and then cut segments off horizontally. Another skill was "fish gill-cutting skill", which was to cut vertically the front of the kidney or squid, with a depth of four-fifths of the thickness of the squid. Then slice horizontally without slicing through for the first cutting and then slice it through, with the even number of cuttings to the end, and then blanch in the boiling water, just like fish gills.

The boiled kidney was a cold dish. Cut the kidney into "fish gills-shaped kidneys" and blanch them. After flushing in cold water, put them in a bowl with lettuce on it; then take the plate to cover and turn over. A large plate looked like kidneys and pours sesame oil, soy sauce and vinegar on it before serving. The "fine eyebrow-like kidney shreds" was another cutting method, which was to slice horizontally and cut vertically.

In Li Qigui's view, Jia Hewu belonged to the "borned chefs", so he taught him very carefully, not only the process of preparing main ingredients, but also the usage of the ingredients in details. One of the basic requirements of cooking was that the cutting methods should be consistent in the same dish, such as shredding, slicing, dicing and blocking and the ingredients couldn't be greater than the main ingredients; otherwise the presumptuous guest usurped the host's role. Whether this served dish

配料，要切成"剔香棍儿"，切得八分长，不到一寸，粗细和剔香的小棍儿差不多。过去老师傅讲，牲口都不吃寸草，所以做菜的主料配料都不能超过寸段儿，就是八分段儿。

要做醋椒鱼，讲究用"蛾眉葱"。把葱剖开，去掉心儿，斜着切叫"蛾眉葱"。"葱烧海参"里的葱也算配料，一般海参从中间一刀，叫"一改二"，个小的不改刀，个长的可以中间斜着抹两刀，这叫抹刀片，基本上都要在一寸五左右，这根据海参的大小和肉的厚度而定。把海参凉水下锅，开后搁上盐，开了锅之后搁料酒，借着料酒热气的挥发把腥味儿带走，然后再煨制，一步一步地做海参。最早做葱烧海参的葱是"蓑衣葱"，葱两面使斜刀，再切成一寸五长的段儿下锅炸。这种花刀切法可以让热油接触到葱心里头，熟了以后夹起来咬的时候，汁水不会滋出来。后来为省事不切蓑衣葱了，直接炸葱段儿。炸葱段不能手懒，如果你手懒了，它就出阴阳面儿，一面煳了一面生，色浅不均。炸葱段要不时搅动着，油温要合适，保持在五六成热。控净了油还要搁上高汤、白糖、精盐等调味品，上屉蒸一下，葱段也不会滋出汁水。过去菜肴的名称都是以珍贵主原料为主取名，但是"葱烧海参"为什么把葱搁在前面呢？是因为葱本身是风味独特的所在，葱的香味跟海参融为一体，用葱香压住海参的腥味。最后把海参的营养跟葱的香味以复合味的形式呈现出来。那这道菜就是葱香味浓，海参软烂，口味儿咸鲜适度。

就这样，贾河武从一个小学徒经过几十年的奋斗，现在担任了最多时拥有18家店面的北京市功德福饮食集团的董事长。从李启贵大师身上学到的不仅仅是厨艺，还包括做人的道理、高尚的厨德、成为一名优秀的餐饮企业负责人的综合素质。

贾河武听进了师父的话，一手抓管理，一手搞研发。1999年，他的团队五六个人参加了第四届全国烹饪大赛，获得了金银铜牌共计11块。这在北京市乃至全国范围内，都算是取得了最佳成绩。

或许人们并不知道，李启贵大师收徒的标准近乎苛刻。像贾河武这样一名优秀的厨师兼餐饮企业家，从20世纪80年代初就跟李大师学艺，但两人真正明确师徒名分，却是在2011年。路遥知马力，日久见人心，这漫长的三十年，既是师徒情分的试金石，更是中华烹饪文化薪火传承的接力棒。

到2000年，在日本新高轮王子饭店举行第三届世界烹饪大赛，李启贵大师找到贾河武说："河武啊，我有个提议，去年全国烹饪大赛你的功德福餐饮有

looked good or not, or harmonious or not, had a lot to do with the matching of main ingredients and ingredients. For example, scallion could be used as main ingredients and ingredients. When scallion was used as an ingredient, it should be cut into "small sticks", which should be about 2.67cm long, with the thickness of about that of small fragrant sticks. In the past, the old master said that animals did not eat inch-long grass, so the main ingredients and ingredients for cooking should not exceed inch-long sections, that's, about 2.67cm long.

To cook Fish with Vinegar and Pepper, "fine eyebrows-like scallion" should be used. Cut the scallion open, remove the core, and cut it obliquely, which is called "fine eyebrows-like scallion". In Scallion Roasted Sea Cucumbers, the scallion was used an ingredient. Generally, the sea cucumber should be cut from the middle, called "cutting one into two". In case of small one, it's not necessary to cut in the middle, and for the long one, it should be cut obliquely in the middle, which was called slicing, basically about 1.5 inch, depending on the size of the sea cucumber and the thickness of the meat. Put sea cucumbers into the cold water of the pot, add salt and cooking wine when the water was boiling, take away the fishy smell with the volatilization of the hot air of cooking wine; then simmer it step by step. The earliest scallion used to cook Scallion Roasted Sea Cucumbers was "straw rain cape-like scallion", cutting obliquely the scallion into 1.5 inch long pieces on both sides and fried in a pan. This cutting method could allow hot oil come into contact with the scallion core, and the sauce would not come out when it was cooked and bitten. Later, in order to save trouble, fry scallion segments directly instead of cutting straw rain cape-like scallion. In frying scallion segments, you should be carefully; otherwise it would have two sides different, that's, one side was burnt and another side was raw with light and uneven color. In frying scallion segments, it's essential to stir from time to time, and the oil temperature should be appropriate and kept at the temperature at 50% to 60% of its boiling point. After the oil was drained, condiments such as stock soup, white sugar and refined salt should be added and steamed together with the main ingredients in a drawer. The scallion segments would not produce sauce. In the past, the names of dishes were all based on precious main raw ingredients, but why did Scallion Roasted Sea Cucumbers put scallion in front? It was because scallion itself had unique flavor. The fragrance of scallion was integrated with sea cucumbers, and the fishy smell of sea cucumbers was suppressed with the fragrance of scallion. Finally, the nutrition of sea cucumbers and the fragrance of scallion were presented in the form of compound flavor. That's to say, this dish was rich in fragrance of scallion, soft and tender in sea cucumbers, and moderate in salty and fresh taste.

In this way, Jia Hewu, after decades of striving, become now the chairman of Beijing Gongdefu Catering Group from a small apprentice, owning at most 18 restaurants. What he learned from Master Li Qigui was not only cooking skills, but also the truth of life, noble cooking ethics and the comprehensive quality of becoming an excellent person in charge of a catering enterprise.

Jia Hewu followed his master's words and engaged in management and research and development at the same time. In 1999, five or six members of his team took part in the 4th National Cooking Competition and won a total of 11 gold, silver and bronze medals, which was the best result in Beijing and even in the whole country.

Perhaps it's unknown that Master Li Qigui followed high-demanding standard for accepting apprentices. An excellent chef and catering entrepreneur like Jia Hewu had studied with Master Li since the early 1980s, but it was in 2011 that the two really defined their relationship of master and apprentice. Time tries all. These long 30 years are not only the touchstone of their relationship, but

限公司已经荣登榜首，但是跟你自己亲自赴赛场一显身手还是不完全一样。赢得了荣誉，也是你的领导能力的体现，可以更加服众。这话对吗？"贾河武听李大师这么一说，沉思了一会儿，抬起头来郑重其事地说："您说得对，我要亲自参赛。"

　　为了一战成功，李启贵和贾河武事先下足了功夫。通过反复推敲，定下了两个主要的参赛菜品：炸空心龙虾球、扒燕菜卷。李启贵大师真是手把手地教，当时正赶上冬季天黑得早，每天晚上八点，李大师在米市大街的天伦王朝酒店下班，亲自开车一个半小时到良乡。贾河武此时已经把厨房收拾干净，等着师父来。爷儿俩连教带学一直到夜里一点半。深夜两点，李启贵再从房山开车回来，将近一个月，天天如此。根据大赛组委会的要求，从原料的准备到原料的粗加工，再到原料的加热烹制，李启贵耐心教，贾河武虚心学。师父做一遍，贾河武做一遍；然后李启贵再做一遍，贾河武再学一遍。"把表挂在墙上，严格按照比赛规定的时间做了四遍，我们俩再一起找出不足的地方，怎么调味？什么火候？用什么样的盘子装？用什么东西点缀？这些都说完了，他来一遍，然后我再来一遍，他再来一遍。这一天的教学宣告结束。"李启贵说："大赛要求90分钟完成两道菜，我就给贾河武掐表，一定要提前10分钟完成，这样才能做到胸有成竹。"

　　"我们天天做，熟能生巧，巧能出精，精到伸手就来。比方说制作空心龙虾球，要用到猪皮，去油去毛之后，把它剁成寸条儿，开水焯过，捞出过凉，然后放上葱姜上屉蒸。看似不难，其实暗藏难点。这个过程贾河武就训练了溜溜两天。难点在哪儿呢？如果你就是开了焯一下捞出来，和多开了两个开儿，肉皮的硬度出来是不一样的，鲜度也不一样。或者多煮了一会儿，同样的水，同样的肉皮，再蒸出来的肉皮冻硬度不够了。贾河武第一次焯的时候就火大了，肉皮冻发软，浓度、硬度和鲜度都没了。第二次又弄这个，还没有达到应有的硬度。第三次李大师亲自上手做，猪皮入沸水，两三秒钟就捞出来，一过凉，上屉再蒸，蒸的时间也不一样。蒸的时间得到什么程度？我就跟他讲，蒸的时候要搁上纯净水，要搁上葱姜，要捏上胡椒粉。然后要蒸三个小时到三个半小时，你前两次蒸两个小时，时间不够。要蒸到什么程度呢？拿竹筷子一夹，啪，这肉皮就断了。这才算成功了。"

　　肉皮冻准备好了，面包渣怎么准备？要用刀切，面包渣不能是软的，因为龙虾的蓉泥是软的，如果面包渣儿也软，就扎不到龙虾茸泥里边去，而是浮

also the baton for the inheritance of Chinese cooking culture.

In 2000 when the 3rd IKA was held in the Grand Prince Hotel Shin Takanawa in Japan, Master Li Qigui found Jia Hewu and said, "Hewu, I have a proposal. In last year's National Cooking Competition, your company Beijing Gongdefu Catering Co., Ltd already topped the list, but it is not exactly the same as your own performance in the competition. Winning the honor is also the embodiment of your leadership ability and would be more convincing to the public. Is this right?" Jia Hewu listened to Master Li's remarks, pondered for a moment, raised his head and said solemnly, "You are right. I would like to take part in the competition personally."

In order to win the competition, Li Qigui and Jia Hewu made great efforts in advance. After repeated deliberation, they finalized two main dishes: Fried Hollow Lobster Balls and Stewed Swallow Rolls. Master Li Qigui really taught him hand in hand. It happened to be in winter, so it was getting dark early. At 8 o'clock every night, Master Li went off work at Sunworld Dynasty Hotel on Mishi Street and drove to Liangxiang for an hour and a half. Jia Hewu had already cleaned up the kitchen and was waiting for Master. He studied under the instruction of Master Li until 1:30 at night. At two o'clock in the night, Li Qigui drove back from Fangshan, which lasted every day for nearly a month. According to the requirements of the organizing committee of the competition, Li Qigui taught patiently and Jia Hewu learned modestly from the preparation of main ingredients, to the rough processing of main ingredients and to the heating and cooking of main ingredients. Master Li did it and Jia Hewu followed him. Then Li Qigui did it again and Jia Hewu learned it again. "We hang the watch on the wall and do it four times in strict accordance with the time stipulated in the competition. We then find out the deficiencies together such as how to season, what temperature, what kind of plate to serve, and what decoration. When all is done, he does it again, I would do it again, and then he will do it again. Then the teaching of the day would be ended." Li Qigui said, "The competition requires chefs to cook two dishes within 90 minutes, so I would reckon by time for Jia Hewu. It's necessary to finish it 10 minutes in advance, so as to be confident."

"We practice every day. Practice makes perfect, which makes proficient, which makes it easy. For example, to make hollow lobster balls, it's essential to use pigskin. After fat and hair are removed, it is chopped into one-inch strips, blanched in boiling water, taken out and cooled, and then steamed in a drawer with scallion and ginger. It seems not difficult, but in fact there are hidden difficulties. Jia Hewu practiced this process for two days. What are the difficulties? If you just blanch it and take it out, the hardness and freshness of the skin would be different from blanching several times. If boiling longer, the steamed pigskin jelly is not hard enough even in the same water and with the same skin. Jia Hewu blanched it under higher temperature for the first time, so the pigskin jelly was soft, without proper concentration, hardness and freshness lost. He failed again in the second time and the pigskin jelly didn't reach the hardness required. For the third time, Master Li made it in person, put the pigskin into boiling water and took out in two or three seconds. Once it was cold, it was steamed in a drawer. The steaming time was also different. "How long should it be steamed? I told him that in steaming, it should be steamed over purified water, with scallion, ginger and pepper. Then it would take three to three and a half hours to steam. You steamed for two hours before, which was not enough. To what extent should it be steamed? Take a pair of bamboo chopsticks to clamp, and it's ok if the pigskin breaks."

The pigskin jelly got ready, how to prepare the bread crumbs? To cut with a knife, the bread

在表面上，下锅一炸，面包渣就都掉了。怎么解决这个问题呢？李启贵给他出了一个主意：你把面包搁到冷库里冻上，冻完之后，把它边上的皮去掉，然后切成薄片，再切成条，顶刀切成丁，称之为面包粒，比麦子粒大，比玉米粒小，放在通风的地方自然风干。风一吹，水汽都走了，拿手一抓，哗啦哗啦的，用这样的面包渣儿包空心龙虾球，抓起一把轻轻一揉，这个面包粒全都均匀地镶入龙虾蓉泥当中。就像赵州桥的石砖一样，相互挤压，浑然一体。这样炸出来的虾球，第一达到了外酥，第二达到了里嫩。虾球就像桥洞一样，外边这个粒挤着那个粒，既是成功之根基，也是成功之美。这个菜体现了力学的特点。

最后到了下锅炸虾球的环节，贾河武又遇到了难题：炸完的虾球颜色不一致，有的偏黑，有的偏红，有的偏黄。还是李大师一语中的，指出了其中的问题所在："你一个一个地往锅里下油炸，第一个虾球到最后一个虾球的受热情况肯定不一样，颜色自然也不一样。这和每一个龙虾球在油炸中耗去的水分息息相关，其嫩度、观感、口感存在差异性。"

李大师的制作方法是，当油温升到四五成热的时候，把虾球一个个捡到漏勺里，同时下入，先不摇动，让温度稍微上一上，虾球一收缩，外皮利落了，就从油的一侧慢慢推起，虾球在锅里不停地旋转。炸至奶金黄色、外焦里嫩。上桌的时候用餐刀切开，汤鲜味浓，外表金黄，称之为上品，方算成功。经李启贵这样精彩演示，贾河武才恍然大悟，掌握了炸空心龙虾球的秘诀。

2000年初，在日本新高轮王子饭店，李启贵作为3个队的领队（既是北京大董烤鸭店的领队，也是北京功德福餐饮有限公司的领队，还是北京全聚德集团的领队），带领这三个队均获得了金牌，其中成绩最好、最值得骄傲的，是贾河武。

贾河武制作完成的这两个菜，博得了世界各国评委的一致好评，夺得了个人赛第一名和团体赛第一名，20年过去了，这个成绩至今无人打破。为什么能取得如此之大的成绩呢？或许背后付出的艰辛只有李启贵和贾河武师徒俩知道。

高光军是李启贵大师的另一位高徒，以前是沈阳军区的一名部队厨师。在李启贵大师的精心栽培下，他在2008年举行的第六届世界烹饪大赛上成为冠军人物。第六届世界烹饪大赛在北京的世纪金源大饭店举行。他参赛的两个菜，一个是葱烧万寿参，一个是炸空心龙虾球。

crumbs couldn't be soft, because the mashed lobster mud was soft. If the bread crumbs were also soft, they couldn't penetrate into the mashed lobster mud, but float on the surface. Once fried in the pan, the bread crumbs would all fall off. How to solve this problem? Li Qigui gave him an idea, "You put the bread in the cold storage and freeze it. After freezing, remove the surface on its edge, cut it into thin slices and then into strips, and then cut it into cubes, which were called bread grains, larger than wheat grains and smaller than corn grains. They should be naturally dried in a ventilated place. As soon as the wind blew, the water vapor had gone away. Grabbed, they crashed. Wrap hollow lobster balls with such bread crumbs, grab a handful and rub them gently, so that bread crumbs were all evenly inlaid into mashed lobster mud. Just like stone bricks in Zhaozhou Bridge, they squeezed each other and became one integrated mass. The shrimp balls fried in this way were tender with a crispy crust. Shrimp balls were just like bridge openings, with crumbs outside squeezing each other, which was not only the essence of success, but also the beauty of success. This dish embodied the characteristics of mechanics."

At last, Jia Hewu encountered another difficult problem in the process of frying shrimp balls in the pan: the fried shrimp balls were of different colors, some were black, some were red and some were yellow. Master Li hit the mark and pointed out the problem, "If you fry the first shrimp ball into the pan one by one, the heating situation from the first shrimp ball to the last shrimp ball will definitely be different, and the color will naturally be different. This is closely related to the water consumed by each lobster ball in frying, and thus its tenderness, appearance and taste are different."

Master Li's processing method was as follows: pick up the shrimp balls one by one into the colander when the oil temperature rose to 40% to 50% of its boiling point, and put them all in the pan at the same time without shaking them until the temperature rose slightly. As soon as the shrimp balls contracted and the skin was loose, pushed them up from one side of the oil, and kept the shrimp balls rotating in the pan until they turned to be golden yellow, tender with a crispy crust. When serving the table, cut them open with a knife. The dish was fresh in soup, delicious in meat and golden in appearance, which reached top grade and thus was considered successful. After such a wonderful demonstration by Li Qigui, Jia Hewu suddenly realized and mastered the secret of frying hollow lobster balls.

At the beginning of 2000, Li Qigui was the leader of three teams (Beijing Dadong Roast Duck Restaurant, Beijing Gongdefu Catering Co., Ltd. and Beijing Quanjude Group) in Grand Prince Hotel Shin Takanawa in Japan. All three teams won gold medals, of which Jia Hewu was the most proud one with the best result.

The two dishes made by Jia Hewu had won unanimous praise from judges all over the world, winning the first place in the individual competition and the first place in the team competition. Twenty years had passed, and no one had broken this record so far. Why could he achieve such great results? Perhaps only Li Qigui and Jia Hewu knew the hardships behind them.

Gao Guangjun was another senior apprentice of Master Li Qigui and used to be an army chef in Shenyang Military Area Command. Under the careful cultivation of Master Li Qigui, he became the champion in the 6th IKA in 2008, which was held at Empark Grand Hotel Beijing. The two dishes he cooked in the competition was Braised Longevity Sea Cucumbers with Scallion and Fried Hollow Lobster Balls.

葱烧万寿参这道菜，源于王义均大师传授给李启贵的丰泽园看家菜"葱烧海参"，经过李启贵大师多年的摸索和创新，推出了这道推陈出新的"葱烧万寿参"。在盘子上手工雕刻了一圈南瓜"寿"字，再配上翠绿的油菜心和煨过的红彤彤的枸杞，上锅蒸熟后浇汁儿，然后把"葱烧海参"装入盘中，整道菜品红黄黑绿，五彩斑斓。不但菜品美观高档、赏心悦目，还在寿宴中体现了喜庆亮丽的主题，最主要的是体现了养生的理念。这道菜既吃到了海参，也吃到了大葱，还吃到了南瓜、油菜心和枸杞，达到了荤素搭配，而且南瓜还有调控血糖过快升高的作用，油菜心清新爽口，枸杞子养生补气。

高光军则是在天伦王朝酒店实习的时候，得到李启贵大师真传的。为了做好这道菜，师徒俩用上了真功夫。李启贵大师亲自开车，带着高光军到大钟寺菜市场、红桥市场、岳各庄菜市场、新发地蔬菜批发中心采购原材料，看看哪儿的龙虾好，哪儿的辽参好，哪儿的大葱好，哪儿的南瓜好。不是有个北瓜倭瓜就能用，瓜的老嫩度以及瓜肉的厚度都有具体的要求。李大师亲自教高光军雕"寿"字，要求他一分钟雕一个，"寿"字要求大小一样，薄厚一样，颜色要统一。

在市场上选辽参，他们选一斤在60头到70头之间的淡干辽参，颜色要一致。李启贵亲自教给徒弟发辽参，师徒二人接力发辽参，然后做对比。发海参首先要洗，洗完了要泡，泡完了要煮一火，然后保温纸封上要焖，焖完了之后要根据海参的软硬度挑选。选好的海参要用冰镇。最后使用的时候，要把海参的韧带去掉，把泥沙洗去，用汤煨好，开始烧制，最后把南瓜"寿"字刻好。菜心处理完了，该蒸的蒸，该煨的煨，达到热度，味都入好了，把海参烧好了放入盘中。这道展现精湛厨艺、色香味形俱佳、健康养生的"葱烧万寿参"，在百年奥运之际，为高光军赢得了第六届世界烹饪大赛的金爵奖。金爵奖就是世界大赛一个国家只有一块纯金加大的金牌。

在颁奖典礼的现场，组委会刚开始宣布成绩，世界烹饪联合会会长杨柳就朝李启贵大师走过来。他兴奋地说："李大师，祝贺您！您的高徒又荣获了世界最高级金奖！"李启贵忙幽默地答道："那要谢谢会长举办的这届比赛，如果没有这次大赛，徒弟们去哪儿展示才华拿金奖呢？"言罢两人哈哈大笑。后来，高光军回到部队，还因此荣立了特等功。

The dish Braised Longevity Sea Cucumbers with Scallion was originated from the specialty of Fengzeyuan Hotel," Braised Sea Cucumbers with Scallion" taught to Li Qigui by Master Wang Yijun. After years of exploration and innovation, Master Li Qigui launched this innovative dish "Braised Longevity Sea Cucumbers with Scallion". A circle of "longevity" character in Chinese was carved in pumpkin by hand on the plate, accompanied by green rape and red simmered Chinese wolfberry. After steaming in a pan, pour the sauce, and then place the" Braised Sea Cucumbers with Scallion" into the plate. The whole dish was colorful, red, yellow, black, and green. It's not only beautiful and high-graded, pleasing to the eye, but also festive and colorful for the birthday party. What's most important was the concept of health preservation. This dish contained sea cucumbers, scallion, pumpkin, rape and Chinese wolfberry, thus achieving the good matching of meat and vegetable. Moreover, pumpkin also has the function of regulating the excessive rise of blood sugar. Rape is fresh and refreshing, and Chinese wolfberry is good for health and nourishes vitality.

Gao Guangjun got the authentic skills of Master Li Qigui when he took internship at Sunworld Dynasty Hotel. In order to cook this dish well, they made great efforts. Master Li Qigui drove in person and took Gao Guangjun to Dazhongsi Vegetable Market, Hongqiao Market, Yuegezhuang Vegetable Market and Xinfadi Vegetable Wholesale Center to purchase main ingredients, and look for good lobsters, sea cucumbers from Liaoning, scallion and pumpkin. They picked up main ingredients strictly in terms of the tenderness and thickness. Master Li personally taught Gao Guangjun to carve the Chinese character "longevity" and asked him to carve one in a minute. The Chinese character "longevity" should be carved in the same size, thickness and color.

When choosing sea cucumbers from Liaoning in the market, they chose a half kilo of dried sea cucumbers from Liaoning involving 60 and 70 heads, with the same color. Li Qigui personally taught his apprentice to water fat sea cucumbers from Liaoning. They relay to the water fat sea cucumbers and then make a comparison. In terms of water fat sea cucumbers, wash them first, soak after washing, boil after soaking, and then seal with the heat preservation paper for standing. After standing, the sea cucumbers must be selected according to the hardness. The selected sea cucumbers should be chilled. When using, the ligaments of sea cucumbers should be removed and the silt should be washed away. Sea cucumbers should be simmered in soup, and then be cooked. Finally, when the Chinese character "longevity" was carved on the pumpkin, the rape was cleaned and everything got done, place the cooled sea cucumber into the plate. Gao Guangjun won the Golden Jazz Award in the 6th IKA on the occasion of the Centennial Olympics with "Braised Longevity Sea Cucumbers with Scallion", which showed his exquisite cooking skills, excellent color, aroma and type, and health reservation. The Golden Jazz Award is a gold medal that a country has only one enlarged gold mental in the world competition.

At the scene of the award ceremony, when the organizing committee had just begun to announce the results, Yang Liu, president of the World Association of Chinese Cuisine, came towards Master Li Qigui. He said excitedly, "Master Li, congratulations! Your senior apprentice has won the world's highest gold medal again!" Li Qigui replied hastily and humorously, "Thank the president for hosting this competition. Without this competition, where would he show his talent and win the gold medal?" Then they burst out laughing. Later, Gao Guangjun returned to the army and was awarded a special merit hereby.

2017年8月3日，李启贵大师、中国烹饪协会专家组到渭南市，考评洽阳黄河生态美食名城并任组长。

On August 3, 2017, Li Qigui headed a group of experts from the China Culinary Association to Weinan City to evaluate a city of Yellow River ecological cuisine in Qiayang.

2019年，在《技艺传承》节目烹制名菜。

In 2019, Li Qigui cooked famous dishes on the program "Skill Inheritance".

2012年2月，在175个国家大使迎新春各国大使招待会上，李启贵主理了中国八珍宝鼎宴。
In February 2012, Li Qigui hosted the Chinese Eight Delicacies Banquet for 175 ambassadors to celebrate the Chinese New Year.

2019年10月，李启贵大师传授徒弟烹饪技艺。
In October 2019, Li Qigui taught his apprentices cooking skills.

李启贵大师与部分徒弟合影。
Group photo of Li Qigui and some of his apprentices.

经典菜品

Classic Dishes

中华八珍宝鼎
Chinese Eight Treasures Tripod

主料：水发鱼翅250克、鲍鱼250克、水发裙边250克、水发海参250克、水发鱼肚200克、水发羊肚菌100克、松茸100克。
配料：生菜100克、金瓜50克、葱5克、姜5克。
调料：清汤1000克、水淀粉150克、盐5克、葱姜油15克、料酒5克、食用油15克、蛋清35克、干贝100克、鲜贝100克。

Main ingredients: 250g water-fat shark's fins, 250g abalones, 250g water-fat turtle rim, 250g water-fat sea cucumbers, 200g water-fat fish maw, 100g water-fat morel and 100g tricholoma matsutake.
Ingredients: 100g lettuce, 50g pumpkin, 5g scallion and 5g ginger.
Seasonings: 1000g clear soup, 150g water starch, 5g salt, 15g fried scallion-ginger oil, 5g cooking wine, 15g edible oil, 35g egg white, 100g dried scallops and 100g fresh scallops.

制作方法 / Steps

1. 将水发鱼翅洗净，加入葱姜料酒蒸透，漂去腥味，再用清汤入味煨透，把鲍鱼制好片片，野生羊肚菌和松茸洗净发透，改刀片片。
2. 海参、裙边、鱼肚，片成抹刀片。
3. 鲜贝去筋，用蛋清、淀粉浆好滑透；干贝去筋洗净，加入清汤蒸好备用。
4. 鼎中放入葱姜油，下入生菜、清汤和以上原料，盐等调味料用水淀粉勾薄芡入锅中，煨至汁浓味厚时加入干贝，用旺火烧开，放入鱼翅即可。

1. Wash water-fat shark's fin with water, add scallion, ginger and cooking wine, steam thoroughly, remove floating foam, simmer thoroughly with clear soup, wash and soap. abalones, wild morel and tricholoma matsutake, and cut them into pieces.
2. Slice water-fat turtle rim, water-fat sea cucumbers and fish maw obliquely.
3. Remove tendons from fresh scallops and smooth with egg white and starch slurry. Remove tendons from dried scallops, wash them, add clear soup and steam them for later use.
4. Add fried scallion-ginger oil into the tripod, add lettuce, clear soup, the above raw materials and seasonings, thicken with water starch, add dried scallops when the sauce becomes thick, and add shark's fins when it's boiling over strong fire.

特点：集八珍原料于一鼎，汤鲜味美，回味无穷。
Features: Integrating eight rare raw materials in a tripod, the dish has delicious soup and endless aftertaste.

北京灵芝烤鸭
Beijing Roast Duck with Ganoderma Lucidum

主料：北京填鸭1只、灵芝35克。
配料：葱15克、黄瓜15克、萝卜15克、酱瓜15克、泡菜15克、荷叶饼适量。
调料：蒜泥5克、白糖5克、甜面酱40克。

Main ingredients: One Beijing stuffed duck and 35g ganoderma lucidum.
Ingredients: 15g scallion, 15g cucumber, 15g radish, 15g pickled melon, 15g pickled vegetable and some cakes.
Seasonings: 5g mashed garlic, 5g white sugar and 40g sweet flour sauce.

制作方法
Steps

1. 选用正宗北京填、鸭经过宰杀、烫坯、开生、晾坯、打色等二十多道工序，方可进行烤制。
2. 在烤鸭时采用外烤内煮的方法，将事先调制好的灵芝水灌入鸭坯体内，烤制过程中，灵芝的营养成分让烤鸭得以充分吸收，片鸭可分为传统片片、片条或皮肉分开三种片法。
3. 烤鸭的专用酱料按比例加入灵芝，荷叶饼在和面时加入灵芝，但要注意面不能太软。
4. 烤鸭片好后与灵芝饼、灵芝甜面酱、酱瓜、泡菜一同上桌即可。

1. The authentic Beijing stuffed duck can only be roasted after more than 20 processes such as killing, blanching, cut-opening, air drying and coloring.
2. When roasting duck, the method of roasting outside and boiling inside is adopted, and the pre-prepared ganoderma lucidum water is poured into the duck. During the roasting process, the nutritional components of Ganoderma lucidum should be fully absorbed into the roasted duck, and the roasted duck can be sliced with three methods: traditional slicing, slicing strips or separating skin from the meat.
3. The special sauce for roasted duck is added with ganoderma lucidum in proportion. The cake is added with ganoderma lucidum when kneading dough, but the dough should not be too soft.
4. Serve the roasted duck with ganoderma lucidum cake, ganoderma lucidum sauce, pickled cucumber and pickles.

特点：外形美观、皮脆肉嫩、酥香不腻、滋补有方。
Features: This dish is beautiful in shape, with crisp skin and tender meat of the roasted duck. It is not greasy but nourishing.

灵芝烤鸭 香酥不腻

▲ 1986年，中国首次参加第五届奥林匹克世界烹饪大赛，此菜荣获世界金奖、金牌证书、金牌纪念杯。
In 1986, China participated in the 5th IKA/Culinary Olympics for the first time and won the gold medal, gold certificate and gold commemorative cup.

一品芙蓉
First-rank Hibiscus

主料：南豆腐250克、鲜贝100克、水发海参100克、大虾肉100克、鲍鱼100克、鲜蘑150克、水发竹荪100克、草菇80克、鸡里脊150克、冬笋80克、黄瓜125克、香菜45克、南荠20克、枸杞20克、油菜心22棵。

调料：葱10克、姜10克、火腿10克、鸡蛋清30克、胡萝卜50克、水发香菇50克、清汤400克、葱姜油15克、精盐5克、料酒5克、水团粉25克、胡椒粉2克、白糖2克、味精3克。

Main ingredients: 250g southern tofu, 100g fresh scallops, 100g sea cucumber, 100g prawn meat, 100g abalone, 150g fresh mushrooms, 100g bamboo fungus, 80g straw mushrooms, 150g chicken fillet, 80g winter bamboo shoots, 125g cucumber, 45g cilantro, 20g nanqi, 20g lycium chinensis, 22 rapeseed shrubs.
Seasonings: 10g green onion, 10g ginger, 10g ham, 30g egg white, 50g carrot, 400g clear soup, 15g oil of green onion and ginger, 5g refined salt, 5g cooking wine, 25g water starch, 2g pepper, 2g sugar, 3g monosodium glutamate.

制作方法
Steps

1. 将油菜心洗净，根部切十字花刀，枸杞发好、镶在菜心上、焯透入味后备用。
2. 将南豆腐、鲜贝、鸡里脊去筋、过箩，加入鸡蛋清、盐、味精、料酒、白糖、水团粉、清汤、胡椒粉调味打匀备用。
3. 将鲍鱼、海参、虾肉、草菇、冬笋、火腿、南荠均切小粒加葱姜制成馅备用。
4. 取一平板模具，将三分之一的蓉泥抹在模具上，中间加入馅料，后再将蓉泥抹平。
5. 将胡萝卜、香菜叶、水发香菇、黄瓜、水发竹荪制成梅、兰、竹、菊、山、鸟的一幅动态图案后、鲜蘑切花刀焯后用清汤煨透入味，码入大盘外圈。
6. 将图案放入屉中蒸熟，拖入大盘中间，摆放整齐匀称。
7. 汤锅上火，放入清汤调好味，点入葱姜油，勾玻璃芡，浇在图案和周围的菜心和鲜蘑上即可。

1. Wash the rapeseed shrubs and cut them crosswise, and set the lycium chinensis on the shrubs, blanch them and set them aside.
2. Blanch southern tofu, fresh scallops, chicken fillet (tendons removed), and add egg white, salt, monosodium glutamate, cooking wine, sugar, water starch, broth, pepper, and mix them evenly.
3. Cut abalone, sea cucumber, shrimp, straw mushroom, winter bamboo shoots, ham and nanqi into small pieces to make the filling and set it aside.
4. Take a flat mold and put a third of the paste on the mold, add the filling in the middle, and then level the paste.
5. Make a dynamic pattern of plum, orchid, bamboo, chrysanthemum, mountain and bird with carrot, cilantro, soaked mushroom, cucumber and soaked bamboo fungus, and cut the fresh mushrooms with superficial cuts and simmer them in broth until they are flavorful, then put them into the outer circle of a large plate.
6. Put the pattern into the drawer to steam and drag it into the middle of a large plate to arrange it neatly and proportionally.
7. Put the soup pot on the fire and add the broth to adjust the taste, add the onion and ginger oil, make crystal-clear gravy and pour it on the pattern, the rapeseed shrubs and fresh mushrooms.

特点：滑嫩鲜香、馅心海鲜味浓、外形美观。
Features: Tender, fresh and fragrant, filled with seafood flavor, beautiful appearance.

▲ 1986年，中国首次参加第五届奥林匹克世界烹饪大赛，此菜荣获世界金奖、金牌证书、金牌纪念杯。
In 1986, China participated in the 5th IKA/Culinary Olympics for the first time and won the gold medal, gold certificate and gold commemorative cup.

百花珍珠鲟

Hundred Flowers Pearl Chinese Sturgeon

主料：鲟鱼1尾1250克、鲜贝14粒。
调料：大虾黄150克、鸽蛋10个、水发香菇15克、香菜5克、豆苗5克、胡萝卜5克、味精3克、盐3克、糖2克、料酒4克、水团粉15克、葱姜15克、清汤适量、姜汁10克、葱姜油少许、鸡蛋清1个。

Main ingredients: 1250g of Chinese sturgeon, 14 fresh scallops.
Seasonings: 150g prawn yolk, 10 pigeon eggs, 15g soaked shiitake mushrooms, 5g cilantro, 5g bean sprouts, 5g carrots, 3g monosodium glutamate, 3g salt, 2g sugar, 4g cooking wine, 1 egg white.

制作方法
Steps

1. 将鲟鱼宰杀洗净后改成瓦垄刀，刀口要均匀，将鲟鱼用盐、味精、料酒腌渍。
2. 鲜贝洗净后去筋，加味精、盐、鸡蛋清、水团粉浆好，然后镶上蒸好的虾黄，镶嵌在鱼的瓦垄刀口上，放入葱姜，上屉蒸熟，放入鱼盘中。
3. 鸽蛋放入勺中，用香菇、香菜、豆苗、胡萝卜装饰出百花图案，上屉蒸熟，码在鱼身两侧。
4. 汤锅上火，放入清汤，烧开后加入盐、料酒、糖、姜汁、水团粉勾芡，加入少量葱姜油，浇于鱼身、百花上即可。

1. Slaughter and clean the sturgeon and cut it in an even manner similar to the rows of roof tile. Make sure the sturgeon is marinated in salt, monosodium glutamate and cooking wine.
2. Wash the scallops and remove the tendons, add monosodium glutamate, salt, egg white, dough powder and mix them well, and then set the steamed shrimp yolk on the cuts, add onion and ginger, and steam them in the drawer, and arrange in the dish.
3. Put pigeon eggs in a spoon. Decorate the pattern with vegetables such as mushrooms. Steam in the upper drawer and arrange then on both sides of the fish.
4. Put the soup pot on the fire and add broth, wait for it to boil, then add salt, cooking wine, sugar, ginger juice and dough powder to make the gravy, add a small amount of onion and ginger oil, pour it on the body of the fish.

特点：鲟鱼营养丰富，蛋白质高，脂肪含量较低，含有饱和脂肪酸，鲟鱼矿物质、维生素较高。
Features: Sturgeon is rich in nutrition, with high level of protein, and low fat content. It contains saturated fatty acids, and high levels of inorganic salts and vitamins.

1986年，中国首次参加第五届奥林匹克世界烹饪大赛，此菜荣获世界金奖、金牌证书、金牌纪念杯。
In 1986, China participated in the 5th IKA/Culinary Olympics for the first time and won the gold medal, gold certificate and gold commemorative cup.

金鱼戏明珠
Goldfish with Pearl

主料：对虾10条1000克。
配料：鸽蛋60克、鸡里脊100克、莴笋100克、胡萝卜20克、黄瓜20克、蛋清15克。
调料：盐4克、料酒3克、葱姜油15克、胡椒粉2克、水淀粉15克、白糖2克、清汤250克、姜汁10克、紫菜3克、樱桃5克、味精2克。

Main ingredients: 10 prawns of 1,000g.
Ingredients: 60g pigeon egg, 100g chicken tenderloin, 100g asparagus lettuce, 20g carrot, 20g cucumber and 15g egg white.
Seasonings: 4g salt, 3g cook wine, 15g fried scallion-ginger oil, 2g pepper, 15g water starch, 2g white sugar, 250g clear soup, 10g ginger juice, 3g laver, 5g cherry and 2g monosodium glutamate.

制作方法
Steps

1. 大对虾洗净，去皮留，尾一分为二，尾部剞十字花刀入味，去虾头斩成蓉，鸡里脊去筋后打成蓉，加入盐、蛋清、料酒、胡椒粉、糖、清汤、水淀粉打匀备用。莴笋切丁后用刻刀雕成八棱空心球，煨制入味。
2. 紫菜剪成丝，黄瓜、胡萝卜切丝，胡萝卜要制成金鱼嘴和眼睛，再将剩余黄瓜和胡萝卜丝切成末，用尺板把蓉抹在大虾尾部，镶上鱼鳞、鱼嘴和眼睛，做成金鱼大虾，鸽蛋煮熟入味。
3. 金鱼大虾上屉蒸熟后，码入盘子周围，中间摆放鸽蛋和莴笋、樱桃。
4. 汤勺上火，放入清汤，烧开后加入盐、味精、料酒、糖、胡椒粉、姜汁，用水淀粉勾薄芡，加入葱姜油，一起调味，浇到金鱼大虾和鸽蛋上即可。

1. Wash and peel prawns, remain the tail and cut in two parts. Cut the tail in a cruciform pattern to taste. Remove the shrimp head and mince the prawns. Mince the chicken tenderloin after removing tendons. Add egg white, salt, cooking wine, pepper, sugar, clear soup and water starch and mix well for later use. Dice asparagus lettuce, carve it into octagonal hollow balls with a carving knife, and simmer to taste.
2. Shred laver, cut cucumbers and carrots, and make carrots as goldfish mouth and eyes. Then mince the remaining cucumbers and carrots. Smear the minced on the tail of prawns with a ruler board, and inlay fish scales, fish mouth and eyes to make goldfish prawns. Pigeon eggs are cooked to taste.
3. After the goldfish and prawns are steamed in a drawer, they are packed around the plate with pigeon eggs, asparagus lettuce and cherries placed in the middle.
4. Add clear soup into the spoon, boil, add salt, monosodium glutamate, cooking wine, sugar, pepper and ginger juice, thicken with water starch, add fried scallion-ginger oil, season together and pour on goldfish, prawns and pigeon eggs.

特点：造型美观、金鱼大虾栩栩如生，虾肉脆嫩多汁。
Features: This dish is beautiful in shape, with lifelike goldfish and prawns, and shrimp meat is crisp, tender and juicy.

茉莉凤蓉竹荪汤

Jasmine, Minced Chicken and Bamboo Fungus Soup

主料：水发天然竹荪45克、鸡里脊150克、清汤1250克。
配料：水发香菇15克、青菜叶15克、胡萝卜25克、茉莉花25克。
调料：水淀粉15克、鸡蛋清15克、糖2克、精盐3克、料酒4克、葱姜油5克。

Main ingredients: 45g natural bamboo fungus, 150g chicken tenderloin, 1250g broth.
Ingredients: 15g mushrooms, 15g green vegetable leaves, 25g carrots, 25g jasmine.
Seasonings: 15g wet starch, 15g egg white, 2g sugar, 3g refined salt, 4g cooking wine, 5g scallion and ginger oil.

制作方法
Steps

1. 将竹荪洗净，泡发好，改刀成菱形，以汤煨透后备用。
2. 将鸡里脊去筋，制成蓉状，加入料酒、精盐、糖、蛋清、水淀粉、葱姜油，加入冷却后的清汤，打匀。
3. 将鸡蓉抹入匙中，用香菇、青菜叶、胡萝卜在上面制成不同形状的花型，蒸透备用。
4. 起锅放入清汤，加入调味料，放入煨透后的竹荪，放入容器内，再放入百花鸡蓉，点缀茉莉花，加盖即可食用。

1. Wash and soak the bamboo fungus and cut them into diamond-shaped slices, simmer them thoroughly with soup, set aside.
2. Remove the tendon of the chicken tenderloin and mince it. Add cooking wine, refined salt, sugar, egg white, starch, scallion and ginger oil and then pour cool broth into it and mix them up.
3. Put the minced chicken into a spoon, make it into different shapes, steam it thoroughly, set aside.
4. Pour broth into the pot, add the seasonings and the simmered bamboo fungus, put them in the container, then put in the minced chicken, garnish with jasmine, cover it with a lid. The dish is ready to be served.

特点：开盖后花香四溢，汤鲜味美，质地软嫩，入口清爽。
Features: After opening the lid, the fragrance of the flowers permeate, the soup is delicious, the mouthfeel is soft and tender, and the taste is fresh.

▲ 1993年，参加"京、沪、川、粤、日本的两国五方大赛"，此菜荣获金奖第一名。

In 1993, it won the gold medal in the "Five-party Competition of Beijing, Shanghai, Sichuan, Guangdong and Japan".

▲ 1993年,参加"京、沪、川、粤、日本的两国五方大赛",此菜荣获金奖第一名。
In 1993, it won the first prize in the "Five-party Competition of Beijing, Shanghai, Sichuan, Guangdong and Japan".

二龙戏珠

Dragon Carrots with Tomato Pearl

主料：酱牛肉150克、盐水虾100克、三文鱼150克、紫菜卷150克、发菜卷150克、五香猪耳100克、冬笋丝150克、糖酥核桃仁100克、糖醋萝卜卷80克、拌西蓝花150克、白蛋糕50克、红色柿子椒15克、小西红柿15克、黑白芝麻5克、墨鱼150克、胡萝卜250克、青笋80克。

调料：食用油25克、精盐8克、白糖6克、料酒6克、大料4克、水团粉10克、酱油15克、葱15克、姜15克、花椒3克、桂皮3克、砂仁2克、丁香1克、肉果3克、醋5克、白芷3克、鸡蛋75克。

Main Ingredients: 150g seasoned beef, 100g salted shrimps, 150g salmon, 150g seaweed roll, 150g hair weeds roll, 100g spiced pig ear, 150g shredded winter bamboo shoots, 100g sugared walnut kernels, 80g sweet and sour radish roll, 150g mixed broccoli, 50g white cake, 15g red bell pepper, 5g black and white sesame, 150g cuttlefish paste, 250g carrots, 80g green bamboo shoots.

Seasonings: 25g cooking oil, 8g refined salt, 6g sugar, 6g cooking wine, 4g star anise, 10g water ball powder, 15g soybean sauce, 15g scallion, 15g ginger, 3g Chinese prickly ash, 3g cinnamon, 2g fructus amomi, 1g cloves, 3g nutmeg, 5g vinegar, 3g radix angelicae, 75g egg.

制作方法
Steps

1. 将以上各种原料，按各种烹调方法加工至成熟，冷却后备用。
2. 将猪耳切片，墨鱼一半切片、另一半卷成1朵花，紫菜卷切成片，牛肉切片，盐水虾一片两开、拌西蓝花切成小朵，萝卜卷切成斜刀卷、冬笋丝做成葱油拌笋丝。
3. 先将冬笋丝码好底，开始第一层摆猪耳，第二层摆墨鱼，第三层放紫菜卷，第四层摆酱牛肉，第五层左侧放盐水虾、中间放三文鱼卷、墨鱼卷、西蓝花，右侧放糖酥核桃仁，第六层左侧摆发菜卷，右侧放糖醋萝卜卷。
4. 把胡萝卜做成龙形，红柿子椒做成火苗形，小西红柿片一刀做成珍珠形，白蛋糕制成白云、青笋做成浪花。
5. 将龙摆放山的左侧一条，在白云中、右下侧摆一条在海浪花中，中间上端是珍珠和火苗，能感觉到栩栩如生。

1. Process and cool the above ingredients according to their various cooking methods, set aside.
2. Slice the spiced pig ear, slice half of the cuttlefish and make the other half into a flower shape, slice seaweed roll and beef, halve the salted shrimps, cut the mixed broccoli into small pieces, cut the radish rolls obliquely, and mix the shredded winter bamboo shoots with scallion oil.
3. Put the shredded winter bamboo shoots at the bottom, put the pig ears at the first layer, the cuttlefish at the second layer, seaweed roll at the third layer, the seasoned beef at the fourth layer, the salted shrimps on the left side, the salmon roll in the middle and the sugared walnut kernels on the right side at the fifth layer, the hair weeds roll on the left side and sweet and sour radish rolls on the right side at the six layer.
4. Make the carrots into dragon shape, red bell pepper into flame shape, small tomato slices into pearl shape, white cake into white clouds and bamboo shoots into spray.
5. Put one dragon carrot on the left side of the layers, one in the middle of the white clouds, one in the low right of the white clouds, one in the spray. With the pearl and flames on the upper end of the middle, this dish is vivid.

特点：拼摆所用原料不同，烹饪方法不同，质地各异，形状美观。
Features: Different ingredients, different cooking methods, diversified texture and delicate appearance.

▲ 1993年，参加全国第三届烹饪大赛，此菜荣获金牌第一名。
In 1993, he won the gold medal in the third National Cooking Competition.

炸空心龙虾球

Fried Hollow Lobster Balls

主料：活龙虾1只3000克。
配料：猪肥膘肉250克、南荠50克、猪皮750克、火腿50克、面包粒75克、葱25克、姜25克、鸡蛋50克、莴笋400克、胡萝卜400克。
调料：清汤750克、盐8克、料酒10克、胡椒粉1克、水淀粉10克、食用油50克、葱姜油25克。

Main ingredients: 1 live lobster of 3,000g.
Ingredients: 250g pig fat meat, 50g water chestnut, 750g pigskin, 50g ham, 75g bread grains, 25g scallion, 25g ginger, 50g eggs, 400g asparagus lettuce and 400g carrot .
Seasonings: 750g clear soup, 8g salt, 10g cooking wine, 1g pepper, 10g water starch, 50g edible oil and 25g fried scallion-ginger oil.

制作方法
Steps

1. 将活龙虾宰杀，头尾要完整，用水加葱姜焯透，控净水分，摆放在盘子的前后两端。龙虾肉去筋皮，与猪肥膘肉、南荠一起均匀切成小粒，放入盆中，加入鸡蛋、盐、胡椒粉、葱姜油、水淀粉，制成虾胶备用。
2. 猪肉皮去净油和毛，切成条状，加入适量清水及清汤，再放入葱、姜、火腿和胡椒粉，上屉将肉皮蒸到软烂即可。然后加入调料调好味，放入焯好的青豆，肉皮过滤，和青豆冷却后制成2.5厘米的方丁。面包去皮切成小粒，将虾胶均匀分成12份，每份包上皮冻，再沾上面包渣制成虾球备用。
3. 勺内放入食用油，待油温达到4~5成热时，下入虾球炸成金黄色捞出，控净油，摆在虾头和虾尾中间，再将胡萝卜和莴笋加工入味，放在盘中，摆放在龙虾两侧即成。

1. Kill the live lobster with the head and tail intact, blanch it thoroughly in boiling water with scallion and ginger, drain water, and place it on the front and rear ends of the plate. After removing tendon and skin, lobster meat is evenly cut into small particles together with pig fat meat and water chestnut, put into a basin, add egg, salt, pepper, fried scallion-ginger oil and water starch, and make minced lobster for later use.
2. Remove the fat and hair from the pigskin, cut it into strips, add clear soup and a proper amount of clear water, then add scallion, ginger, ham and pepper, and steam the pig skin in a drawer until it is soft and tender. Then add seasonings to taste, add blanched green beans, filter the pigskin and cool the green beans to make 2.5cm cubes. Peel the bread and cut it into small particles and divide minced lobster into 12 portions evenly. Each portion is covered with pigskin jelly and then dipped with bread crumbs to make lobster balls for later use.
3. Add edible oil into the pan. When the oil temperature reaches 40-50% of boiling point, add lobster balls, fry them into golden yellow, take them out, drain the oil, place them between the lobster head and the lobster tail, process and taste carrots and asparagus lettuce, and place them on both sides of the lobster.

特点：造型美观、虾球色泽金黄、外焦里嫩、汤汁鲜香。
Features: Beautiful shape, golden color of lobster balls, crisp outside and tender inside, fresh and fragrant soup.

清汤贝蓉翡翠蝴蝶
Green Butterflies with Minced Shellfish in Clear Soup

主料：天然竹荪250克、鲜贝100克、鸡里脊450克、鱼翅针15克。
配料：菠菜叶15克、黄瓜15克、胡萝卜15克、蛋清20克。
调料：盐6克、料酒5克、姜汁10克、糖2克、味精1克、水淀粉15克、胡椒粉1克、葱姜油15克、黑芝麻2克、清汤适量。

Main ingredients: 250g natural dictyophora indusiata, 100g fresh shellfish, 450g chicken tenderloin and 15g shark's fin needle.
Ingredients: 15g spinach leaves, 15g cucumber, 15g carrot and 20g egg white.
Seasonings: 6g salt, 5g cook wine, 10g ginger juice, 2g sugar, 1g monosodium glutamate, 15g water starch, 1g pepper, 15g fried scallion-ginger oil, 2g black sesame and a proper amount of clear soup.

制作方法
Steps

1. 把菠菜叶制成汁备用；鲜贝、鸡里脊去筋制成蓉，加入盐、味精、姜汁、料酒、胡椒粉、蛋清、水淀粉、糖，搅匀后点入葱姜油，分成两份，一份加入菠菜汁备用。
2. 胡萝卜去皮，切成细丝；水发香菇用表皮，切细丝；黄瓜也切成细丝；鱼翅针洗净，用清汤煨制入味备用。
3. 竹荪发好去根，片成片状，控去水分后抹上绿色泥子，蒸制成熟后，中间用白色泥子制成蝴蝶身子，用黄瓜丝、胡萝卜丝点缀，用黑芝麻点缀眼睛，用鱼翅针做蝴蝶须子，上屉蒸透备用。
4. 将白色的泥子挤成扁圆形，放入锅内汆熟备用。
5. 汤勺内放入清汤，调好味，放入白色贝蓉丸子，入味后装入容器内，把蒸熟的蝴蝶均匀地码放在周围即可。

1. Make spinach leaves into juice for later use; remove tendons from fresh shellfish and chicken tenderloin to mince them, add salt, monosodium glutamate, ginger juice, cooking wine, pepper, egg white, water starch and sugar, stir well, add fried scallion-ginger oil, divide into two parts, and add spinach juice to one part for later use.
2. Peel carrots and cut them into fine shreds, cut the upper part of water-fat lentinus edodes into fine shreds, cut cucumbers into fine shreds, wash shark's fin needles and simmer them in clear soup for later use.
3. Water fat natural dictyophora indusiata, remove their roots and slice them; drain water, apply green putty on them and them steam; apply white putty in the middle to make butterfly body, garnish with cucumber and carrot, use black sesame to make eyes, use shark's fin needle to make butterfly beard, and then steam in the drawer for later use.
4. Squeeze the white putty into a flat circle and boil it in a pan for later use.
5. Add clear soup into the soup spoon, add seasonings and white shellfish balls to taste, put them into a container, and evenly stack the steamed butterflies around.

特点：汤鲜味美、清澈见底、贝蓉嫩滑入口即化、造型美观。
Features: The dish tastes delicious with fresh and clear soup, tender and smooth minced shellfish, as well as a beautiful shape.

▲ 1993年,参加全国第三届烹饪大赛,此菜荣获金牌第一名。
In 1993, he won the gold medal in the third National Cooking Competition.

什锦攒盘

Assorted Dishes

主料：莴笋75克、小扁豆50克、黄瓜50克、青豆35克、白萝卜75克、红萝卜75克、鸡里脊400克、牛肉400克、发菜4克、鸡蛋75克、鱿鱼150克、鸭脯肉400克、猪腰子400克、紫菜35克、虾仁100克、猪皮500克、香菜5克、生菜50克、蛋清15克、猪里脊400克。

调料：盐10克、料酒10克、水淀粉35克、香油10克、小西红柿3克、葱25克、姜25克、糖15克、醋15克、辣椒5克、味精2克、黄酱50克、酱油15克、桂皮15克、大料10克、白果10克、丁香3克、白芷5克、砂仁5克、胡椒粉2克、花椒5克、小茴香15克、面粉50克、葱姜油25克、花椒油2克、姜汁2克。

Main ingredients: 75g asparagus lettuce, 50g lentil, 50g cucumber, 35g green bean, 75g white radish, 75g carrot, 400g chicken tenderloin, 400g beef, 4g hair weeds, 75g egg, 150g squid, 400g duck breast, 400g pig kidney, 35g laver, 100g shrimp, 500g pigskin, 5g coriander, 50g lettuce, 15g egg white and 400g pig tenderloin.

Seasonings: 10g salt, 10g cooking wine, 35g water starch, 10g sesame oil, 3g cherry tomatoes, 25g scallion, 25g ginger, 15g sugar, 15g vinegar, 5g pepper, 2g monosodium glutamate, 50g salted and fermented soya paste, 15g soy sauce, 15g cinnamon, 10g Chinese anise, 10g ginkgo, 3g clove, 5g angelica dahurica, 5g amomum villosum, 2g pepper powder, 5g prickly ash, 15g fennel, 50g flour, 25g fried scallion-ginger oil, 2g zanthoxylum oil and 2g ginger juice.

制作方法
Steps

1. 将鱿鱼去皮、打鱼鳞花刀，焯好控净水后加盐、味精、花椒油炝拌，码入盘中。
2. 莴笋去皮，切薄片，焯水过凉，控净水后加香油、盐、味精，拌好以双拼的形式码入盘中。
3. 鸡蛋加面粉吊制成蛋皮，鸡里脊去筋制成蓉，加入料酒、盐、蛋清、水淀粉、胡椒粉、发菜和鸡蓉拌匀，做成发菜卷蒸熟，晾凉后斜刀切成卷，码成花形。
4. 牛肉焯水后，用桂皮、大料、白果、丁香、白芷、砂仁、胡椒粉、花椒、小茴香、黄酱、酱制成五香牛肉，切片码在盘子的一端，小扁豆焯熟后加入姜汁做成姜汁扁豆，码放在牛肉旁边。
5. 猪里脊制成蓉，加入盐、味精、料酒、鸡蛋、水淀粉、胡椒粉，打匀后用蛋皮和紫菜制成卷，切片码入盘中。
6. 将猪腰子去皮，打鱼鳃花刀，用水焯后拌三合油码入盘中，用生菜垫底。
7. 白萝卜切片，用盐、糖腌好，红萝卜切丝，制作成糖醋萝卜卷斜刀切后码入盘中。
8. 取黄瓜皮腌制后去水分，加糖醋，辣椒切丝用热油炸成辣椒油放入盆中拌匀，晾凉后卷成卷；鸭脯制成蓉，加入酱油、料酒、水淀粉、胡椒粉、鸡蛋，拌匀放入容器蒸熟后，切片码入盘中成双拼形状。
9. 猪皮焯水后改刀成条状，加入青豆、虾仁、料酒、盐、胡椒粉、葱姜，蒸熟后制成水晶冻，切成菱形块码放在盘子中间，上面点缀小西红柿即可。

1. Peel squid, cut it into fish scale shapes, blanch them, drain water, add salt, monosodium glutamate and pepper oil, mix well, and put into a plate.
2. Peel, slice, blanch and cool asparagus lettuce, drain water, add sesame oil, salt and monosodium glutamate, mix it well and put it into a plate in the form of double parts.
3. Make egg pastry with eggs and flour, remove tendons from chicken tenderloin to make paste, add cooking wine, salt, egg white, water starch, pepper powder, hair weeds and chicken paste, mix them well, make hair weeds rolls, steam them, cool them in air, cut them obliquely into short rolls, and stock them into flower shapes.
4. Blanch the beef and spice beef with cinnamon, Chinese anise, ginkgo, clove, angelica dahurica, amomum villosum, pepper powder, prickly ash, fennel and other seasonings and sauce; slice the spiced beef and stack at one end of the plate; blanch the lentils, add ginger juice to make ginger juice lentils, and stack beside the beef.
5. Make pork tenderloin into paste, add salt, monosodium glutamate, cooking wine, eggs, water starch and pepper powder, mince evenly, make rolls with egg pastry and laver, slice and stack in the plate.
6. Peel pig kidneys, cut it in a fish gill pattern, blanch it, mix it with three oils and put them into a plate, with lettuce at the bottom.
7. Slice the white radish, marinate it with salt and sugar, shred the carrot, make it into sweet and sour radish, cut the rolls obliquely and put it into a plate.
8. Pickle cucumber skin, remove moisture, add sugar and vinegar, shred chili, fry chili oil with hot oil, put into a basin, mix evenly, cool in the air, and make it into rolls; make duck breast into paste, add soy sauce, cooking wine, water starch, pepper powder and eggs, mix them well, steam in a container, slice and stack it in a plate to form a double-part shape.
9. Blanch pigskin, cut it into strips, add green bean, shrimp, cooking wine, salt, pepper powder, scallion and ginger, steam them, make crystal jelly, cut them into rhombus pieces and put them in the middle of the plate, with cherry tomatoes on top.

特点：刀工讲究、色泽美观、口味各异、营养丰富。

Features: This dish is exquisite in cutting skills, beautiful in color, special in taste and rich in nutrition.

▲ 此菜经北京烹饪协会认定，列为京贵八珍宝鼎宴名菜。
It is a famous dish of Beijing Cuisine Association and Beijing Eight Delicacies Banquet.

▲ 此菜经北京烹饪协会认定，列为京贵八珍宝鼎宴名菜。
It is a famous dish of Beijing Cuisine Association and Beijing Eight Delicacies Banquet.

浇汁海上鲜
Seafood with Sauce

主料：鲜活东星斑一尾1250克。
配料：水发辽参75克、基围虾50克、鲜活鱿鱼50克、葱5克、姜5克、青蒜5克、蒜5克。
调料：食用油125克、水淀粉175克、酱油5克、清汤150克、盐5克、蚝油3克、料酒5克、醋5克、味精2克。

Main ingredients: 1 live plectropomus leopardus of 1,250g.
Ingredients: 75g water-fat sea cucumbers from Liaoning, 50g metapenaeus ensis, 50g live squid, 5g scallion, 5g ginger, 5g garlic sprouts and 5g garlic.
Seasonings: 125g edible oil, 175g water starch, 5g soy sauce, 150g clear soup, 5g salt, 3g oyster sauce, 5g cooking wine, 5g vinegar and 2g monosodium glutamate.

制作方法 / Steps

1. 将东星斑宰杀后，鱼两面均匀改刀成翻刀片；将海参切成抹刀片；鱿鱼去皮洗净，改刀成鱼鳃花刀；基围虾去皮。三种原料用开水焯透备用。
2. 将葱切丝、蒜切片、青蒜切段、姜切成米放入碗中，加入清汤、水淀粉、酱油、盐、料酒、蚝油、醋、味精对成碗汁。
3. 勺中放油烧热，将东星斑两面挂匀水淀粉，用热油炸至外焦里嫩后，捞出控净，油放入盘中。
4. 炒勺上火，放入食用油，烧热后烹入碗汁，待炒出香气后下入海参、鱿鱼、基围虾，炒至成熟后浇到炸好的东星斑上即可。

1. Kill the plectropomus leopardus, cut the two sides of the fish into uniform pieces obliquely, slice sea cucumbers obliquely, remove the peel of the squid, wash it clean and then cut the squid into fish gill-shaped pattern, remove the peel of the metapenaeus ensis, and blanch three raw materials with boiling water for later use.
2. Shred scallion, slice garlic, section garlic sprouts, mince ginger and put it into a bowl. Add clear soup, water starch, soy sauce, salt, cooking wine, oyster sauce, vinegar and monosodium glutamate to make a sauce.
3. Heat up the oil in the pan, thicken the fish with water starch evenly on both sides, fry with hot oil to make the outside crisp and inside tender, then take it out to drain oil and put it into the plate.
4. Add edible oil to the frying pan, heat it and pour the sauce in it. Stir-fry for a short while with the smell of the sauce, add sea cucumbers, squid and metapenaeus ensis, and pour it on the fried plectropomus leopardus after they are cooked well.

特点：东星斑外焦里嫩，海参、鱿鱼、基围虾质地各异，海鲜味浓香，色泽美观。
Features: The plectropomus leopardus is tender with a crispy crust. Sea cucumbers, squid and metapenaeus ensis have different tastes, with strong seafood flavor and beautiful color.

全爆海八珍
Quick-fried Eight Delicacies of Seafood

主料：活龙虾1只2500克、鲍鱼5头10个1000克、水发海参100克、鲜贝100克、水发裙边100克、鲍贝100克、水发鱼肚60克、螺片60克。
配料：葱5克、青蒜5克、老母鸡400克、排骨400克、火腿100克、鸡蛋清30克。
调料：清汤300克、水淀粉40克、姜汁5克、料酒5克、醋2克、盐3克、食用油15克、蚝油3克。

Main ingredients: 1 live lobster of 2,500g, 10 five-head abalones of 1,000g, 100g sea cucumbers, 100g fresh shellfish, 100g water-fat turtle rim, 100g dried abalone, 60g water-fat fish maw and 60g sliced whelk.
Ingredients: 5g scallion, 5g garlic sprouts, 400g old hen, 400g pork ribs, 100g ham and 30g egg white.
Seasonings: 300g clear soup, 40g water starch, 5g ginger juice, 5g cooking wine, 2g vinegar, 3g salt, 15g edible oil and 3g oyster sauce.

制作方法
Steps

1. 将龙虾宰杀后，虾肉改刀成丁，龙虾头尾煮透入味，控净水分，码放在盘子的两端；鲍鱼壳焯水后放在中间；发制好的鲍鱼洗刷干净，老母鸡、排骨改刀成块、炸好，用砂锅加箅子码放一层排骨，上面放一层鲍鱼，再放上老母鸡煲好备用。
2. 把龙虾肉和鲜贝分别用鸡蛋清和水淀粉浆好滑油，其他原料分别改刀切丁，煨制入味，再用热油滑熟；勾兑碗汁，加入清汤、姜汁、料酒、盐、醋、蚝油、葱、青蒜段、水淀粉，烹入锅中爆好，装入鲍鱼壳中。
3. 从砂锅中取出已煲制好的鲍鱼，码放在鲍鱼壳的周围，一边放5个，将勺内鲍汁收，调好，味浇在鲍鱼上即可。

1. Kill and dice the lobster, boil the head and tail of the lobster thoroughly to taste, drain the water and place it at both ends of the plate; blanch abalone shells and put them in the middle; wash the prepared abalone, cut the old hen and pork ribs into large pieces and fry them, put a layer of ribs in the perforated strainer of a casserole, put a layer of abalone on top, and then put the old hen on and cook for later use.
2. Thicken lobster meat and fresh shellfish with egg white and water starch respectively, and fry them in the oil over lower fire; dice other raw materials respectively, simmer to taste, and then fry with hot oil; blend sauce, add clear soup, ginger juice, cooking wine, salt, vinegar, oyster sauce, scallion, garlic sprouts and water starch, stir-fry in a pan, and put into abalone shells.
3. Take out the cooked abalones from the casserole, place it around the abalone shells, place 5 at each side, and pour the thickened and seasoned abalone sauce in the pan into the abalone.

特点：八珍原料质地各异、海鲜味浓、色泽美观、营养丰富。
Features: The dish have different textures of raw materials, strong seafood flavor, beautiful color and rich nutrition.

▼ 李启贵烹制此菜，荣获全国十佳烹饪大师称号。
He won the title of Top 10 Chefs in China with this dish.

葱烧万寿参
Ginseng Stewed with Scallion

主料：水发刺参（长岛参）10条1000克。
配料：章丘大葱60克、油菜心400克、枸杞10克、南瓜500克。
调料：清汤150克、糊葱油75克、盐5克、料酒4克、烧海参汁350克、水淀粉50克、香油适量、食用油15克。

Main ingredients: 10 water-fat sea cucumbers from Changdao (stichopus japonicas) of 1,000g.
Ingredients: 60g scallion from Zhangqiu, 400g rape, 10g Chinese wolfberry and 500g pumpkin.
Seasonings: 150g clear soup, 75g paste scallion oil, 5g salt, 4g cooking wine, 350g braised sea cucumber juice, 50g water starch, 15g edible oil and sesame oil.

制作方法 / Steps

1. 先将葱洗净，选葱白切成六段，每段7厘米长，用油炸至金黄色，加入烧海参汁，蒸制后备用；将南瓜去皮，刻成寿字蒸熟，油菜心根部改成十字花刀；枸杞发好，嵌入油菜心，入味备用。
2. 将发好的海参洗净，放入勺中，加入清汤、料酒、香油、盐、葱油，煨制入味。
3. 炒勺上火，下入糊葱油烧热，放入海参，下入烧海参汁和葱段一起烧制入味，用水淀粉勾芡，烹入料酒、加入清汤、盐、糊葱油，小火慢煨。
4. 将蒸好的寿字、油菜心码放盘子周围，下面放蒸好的葱段，上面浇上白汁，海参煨透，入味勾芡，点入糊葱油出锅，放在盘子的中央，上面点缀煸好入味的青蒜段即可。

1. Wash scallion first, cut scallion into six sections, each 7cm long, fry until golden yellow, add roasted sea cucumber sauce, and steam for later use; peel pumpkins, carve them with Chinese characters longevity and steam them, and cut the root of heart in a cruciform pattern; soap Chinese wolfberry and be embedded into rape to taste for later use.
2. Wash the water-fat sea cucumbers, put them into a pot, add clear soup, cooking wine, sesame oil, salt and scallion oil, and simmer to taste.
3. Heat the frying spoon with paste scallion oil, add sea cucumbers, add sea cucumber sauce and scallion sections, cook them together to taste, thicken with water starch, add cooking wine, clear soup, salt and scallion oil, and simmer slowly over low fire.
4. Place the steamed Chinese characters longevity and rape around the plate, place the steamed scallion sections at the bottom of the plate, pour white sauce on the top, simmer sea cucumbers thoroughly and thicken with water starch, add the paste scallion oil, take it out of the pan, place it in the center of the plate, and embellish with the fried scallion sections on the top.

特点：色泽红亮、葱香浓郁、柔软滑润、造型美观。
Features: The dish is red and bright in color, rich in scallion fragrance and beautiful in shape, and tastes soft and smooth.

此菜荣获第六届中国烹饪世界大赛金奖。
It won the gold medal of the sixth World Cooking Competition.

▲ 此菜荣获中国烹饪协会颁发的首批大师宴名菜证书。
It was granted with the certificate of the first batch of famous dish in the master banquet issued by the Chinese Cuisine Association.

罗汉干燶虾

Quick-fried Prawn

主料：大对虾10条1000克。
配料：水发香菇10克、冬笋10克、面包粒50克、脂油丁15克、豌豆10克、黑芝麻10克、胡萝卜250克、蛋清20克、葱5克、姜5克。
调料：清汤150克、盐3克、葱姜油15克、胡椒粉1克、白糖10克、料酒3克、豆瓣酱20克、淀粉15克、食用油25克。

Main ingredients: 10 prawns of 1,000g.
Ingredients: 10g water-fat lentinus edodes, 10g winter bamboo shoots, 50g bread crumbs, 15g diced fat, 10g peas, 10g black sesame, 250g carrots, 20g egg white, 5g scallion and 5g ginger.
Seasonings: 150g clear soup, 3g salt, 15g fried scallion-ginger oil, 1g pepper, 10g white sugar, 3g cooking wine, 20g broad bean paste, 15g starch and 25g edible oil.

 制作方法
Steps

1. 将大对虾洗净、去沙线、沙包，改刀头尾分开，把虾尾部片开剞花，刀加入调料腌制入味；胡萝卜刻成元宝形状，焯水入味备用。
2. 冬笋、水发香菇、胡萝卜、脂油切成丁，与豌豆一起焯水备用。
3. 虾头部分的虾肉加入肥肉丁搅制成虾泥子，加入蛋清、淀粉、盐、料酒、胡椒粉，制成虾球，镶在虾尾上，沾上面包粒，点缀上黑芝麻，在热油中炸制成熟后，放入盘子的一端。
4. 炒勺上火，放入葱姜油，烧热后下入脂油丁、葱姜、豆瓣酱煸炒，烹入料酒，放入另一半大虾，煸至大虾收缩时下入糖、清汤、盐燶制，加入配料，调好味收汁出锅，放入盘子另一端，中间放胡萝卜元宝，浇上白汁即可。

1. Wash prawns, remove sand line and sand bag, cut off the head and tail, slice the tail of the prawns unbrokenly, add seasonings to taste, carve carrots into shoe-shaped gold ingots, and blanch to taste for later use.
2. Cut winter bamboo shoots, water-fat lentinus edodes, carrots and fat into cubes and blanch them with peas for later use.
3. Add diced fat meat to the head meat of prawns and mince it to make minced prawns. Add egg white, starch, salt, cooking wine and pepper to make prawn balls. Set them on the prawn tail, stick them with bread crumbs, embellish them with black sesame seeds, fry them in hot oil, and then put them on one end of the plate.
4. Add fried scallion-ginger oil to the frying spoon, heat it, stir-fry diced fat, scallion and ginger, bean paste, and cooking wine, add the other half of prawns, stir-fry until the prawns shrink, add sugar, clear soup and salt to simmer, add ingredients to taste, reduce the sauce, put it on the other end of the plate, put shoe-shaped gold ingots of carrot in the middle, and pour white sauce.

特点：一虾两吃、口感独特、鲜美各异。
Features: The dish has two flavors of prawns with unique taste and different delicacies.

▲ 此菜荣获中国烹饪协会颁发的首批大师宴名菜证书。
It was granted with the certificate of the first batch of famous dish in the master banquet issued by the Chinese Cuisine Association.

御鼎鸳鸯燕

Braised Red Cubilose and Fine White Cubilose

主料：官燕25克、血燕5克。
配料：菊花豆腐100克。
调料：清汤250克、盐3克、姜汁5克、料酒3克、碱面3克。

Main ingredients: 25g fine white cubilose and 5g red cubilose.
Ingredients: 100g chrysanthemum tofu.
Seasonings: 250g clear soup, 3g salt, 5g ginger juice, 3g cooking wine and 3g alkaline flour.

制作方法
Steps

1. 将官燕发好，去掉燕毛和杂质，血燕处理方法和官燕相同。
2. 燕窝处理干净后，控净水分放入碗内，加入碱面，冲入开水发制5~6分钟，用筷子搅动，待没有硬心时控去水分，放入碱面，每放一次碱、燕窝就长一倍，要连续撒三四次，直到没有碱味了，控去水，装入盘中，下面是官燕、上面是血燕。
3. 豆腐切成菊花形状，焯水后放入清汤中，以盐、姜汁、料酒煨透入味，先将鼎内放入清汤，再将菊花豆腐放入鼎中即可。

1. Remove the hair and impurities of the fine white cubilose and the red cubilose;
2. Clean cubilose, drain water, put it into a bowl, add alkali flour, pour boiled water into the bowl and soak for 5-6 minutes, stir with chopsticks, drain water when there is no hard core, and add alkali flour; each time alkali is added, the cubilose will double in growth, and add alkali three to four times continuously until there is no alkali smell; drain the water, and put it into the bowl, with fine white cubilose below and red cubilose above.
3. Tofu is cut into chrysanthemum shapes, blanched and simmered in clear soup. Put clear soup into the tripod, and then put the chrysanthemum tofu.

特点：燕窝是滋补佳品，汤鲜味美，脆嫩适口。
Features: The cubilose is a good nourishing product. The dish tastes crisp, tender and palatable with delicious soup.

如意荷包三鲜

Pocketed Shrimps, Sea Cucumber and Chicken Breast

主料：水发海参250克、大虾肉250克、冬笋100克、鸡里脊300克。
配料：胡萝卜250克、鸡蛋300克、青豆10克、红椒丝5克。
调料：清汤250克、料酒3克、胡椒粉1克、水淀粉15克、葱姜油15克、姜汁10克、糖2克、盐5克、食用油适量。

Main ingredients: 250g water-fat sea cucumbers, 250g prawn meat, 100g winter bamboo shoot and 300g chicken tenderloin.
Ingredients: 250g carrot, 300g egg, 10g green beans and 5g shredded red pepper.
Seasonings: 250g clear soup, 3g cooking wine, 1g pepper, 15g water starch, 15g fried scallion-ginger oil, 10g ginger juice, 2g sugar, 5g salt and a proper amount of edible oil.

制作方法
Steps

1. 将鸡里脊去筋后制成蓉，分成两份；海参洗净，焯透，控去水分，和大虾肉、冬笋都切成粒，调入葱姜油做成三鲜馅。
2. 碗内加入鸡蛋、少量水淀粉、盐，勺内放入油上火烧热，将鸡蛋拌匀后吊制成蛋皮，中间放入三鲜馅，用筷子夹住做成荷包，上面点缀红椒丝。
3. 将另一份鸡蓉再分成两份调好味，点缀上青豆和胡萝卜，上屉蒸透后码入盘中，盘子中间摆放已经入味刻好的如意，周围摆放蒸透入味的荷包。
4. 汤勺上火，放入清汤烧开后，加入盐、料酒、糖、姜汁、胡椒粉，最后用水淀粉勾薄芡，浇在如意上即可。

1. Remove the tendons of the chicken tenderloin, make it into paste and divide it into two parts; wash the sea cucumbers, blanch them thoroughly, drain water, cut them with shrimp meat and winter bamboo shoots into grains to make three fresh fillings with fried seallion-ginger oil.
2. Add eggs, a small amount of water starch and salt into the bowl, add oil into the spoon and heat it. Mix the eggs evenly and make it into make egg pastry. Put three fresh fillings in the middle of the egg pastry, and hold them with chopsticks to make a purse, which is decorated with red pepper shreds.
3. Divide the other part of chicken paste into two parts, add seasonings, embellish with green beans and carrots, steam it thoroughly in the drawer and then stack it into the plate. Place the carved Ruyi purse in the middle of the plate, and place the steamed purse around it.
4. Heat the soup spoon, add clear soup and boil it; add salt, cooking wine, sugar, ginger juice and pepper, finally thicken with water starch, and pour on Ruyi purse.

特点：寓意吉祥，荷包口味鲜美、造型美观、色彩丰富。
Features: This dish means auspicious, andis delicious in taste, beautiful in shape and rich in color.

▼ 此菜荣获中国烹饪协会颁发的首批大师宴名菜证书
It was granted with the certificate of the first batch of famous dish in the master banquet issued by the Chinese Cuisine Association.

香菌素烩
Braised Mushrooms

主料：松露60克、榄菜60克、胡萝卜60克、莴笋60克、银耳60克。
调料：清汤500克、盐4克、白糖2克、料酒3克、葱姜油15克、水淀粉15克、鸡油5克、胡椒粉2克、味精2克。

Main ingredients: 60g truffle, 60g olive, 60g carrot, 60g asparagus lettuce and 60g tremella.
Seasonings: 500g clear soup, 4g salt, 2g white sugar, 3g cooking wine, 15g fried scallion-ginger oil, 15g water starch, 5g chicken oil, 2g pepper and 2g monosodium glutamate.

制作方法
Steps

1. 将莴笋、胡萝卜去皮，制成球状，用开水焯好，用清汤煨透入味。
2. 松露洗净后加入清汤，蒸透入味备用。
3. 胡萝卜、莴笋、榄菜焯水时加入少量葱姜油、糖、盐、料酒，然后用清汤煨制入味；银耳发透后去根，放入碗中加鲜汤蒸透入味。
4. 银耳码放在大盘中间，其他原料码在周围，勺中放入清汤、盐、味精、胡椒粉、糖、水淀粉勾白汁，放入少量、鸡油，葱姜油浇在原料上即可。

1. Peel asparagus lettuce and carrot, make them into balls, blanch them with boiling water, and simmer them with clear soup to taste.
2. Wash truffles, add clear soup, steam thoroughly and taste for later use.
3. Add a small amount of fried scallion-ginger oil, sugar, salt and cooking wine when blanching carrots, asparagus lettuce and olive vegetables, and then simmer with clear soup to taste; remove the root of tremella after it is thoroughly soaped, put it into a bowl, add fresh soup and steam until it tastes delicious.
4. Place tremella in the middle of the plate with other raw materials around. Place clear soup, salt, monosodium glutamate, pepper, sugar and water starch to thicken in the spoon, and pour a small amount of fried scallion-ginger oil and chicken oil on the raw materials.

特点：素菜佳肴、造型美观、营养丰富、色彩鲜艳。
Features: This vegetarian dish is beautiful in shape, rich in nutrition and bright in color.

▼ 此菜荣获中国烹饪协会颁发的首批大师宴名菜证书。
It was granted with the certificate of the first batch of famous dish in the master banquet issued by the Chinese Cuisine Association.

▲ 此菜荣获中国烹饪协会颁发的首批大师宴名菜证书。
It was granted with the certificate of the first batch of famous dish in the master banquet issued by the Chinese Cuisine Association.

炒燕窝
Fried Cubilose

主料：水发燕窝500克。
配料：金华火腿15克、鸡蛋清200克。
调料：清汤250克、盐4克、姜汁10克、糖1克、水淀粉15克、葱姜油15克、鸡油5克、碱面3克。

Main ingredients: 500g water-fat cubilose.
Ingredients: 15g Jinhua ham and 200g egg white.
Seasonings: 250g clear soup, 4g salt, 10g ginger juice, 1g sugar, 15g water starch, 15g fried scallion-ginger oil, 5g chicken oil and 3g alkaline flour.

制作方法
Steps

1. 将干燕窝用清水泡透，去除燕毛和杂质备用。金华火腿切成火腿末，放入碗中加碱发透后，除去碱味后用清汤煨制入味，控干水分。
2. 发好的燕窝放碗内，加入清汤、盐、姜汁、火腿末、鸡蛋清、糖、水淀粉，拌匀备用。
3. 炒勺上火，烧热，放入葱姜油、鸡油，然后放入燕窝，用微火慢慢推炒，炒至滑嫩成形时，再点入少量鸡油和葱姜油，装盘即可。

1. Soak the dried cubilose thoroughly in clear water and remove swallow hair and impurities for later use. Cut Jinhua ham into minced ham, put it into a bowl, add alkali to water-fat it thoroughly, remove alkali flavor, simmer with clear soup to taste, and drain water.
2. Add clear soup, salt, ginger juice, minced ham, egg white, water starch and sugar into the bowl and mix them well for later use.
3. Heat the frying spoon over the fire, add fried scallion-ginger oil and chicken oil, then add the cubilose, and slowly fry over low fire until it is smooth and tender; then add a small amount of chicken oil and fried scallion-ginger oil and put them on a plate.

特点：色泽洁白、质地滑嫩、营养丰富、口感软糯。
Features: This dish is white in color, smooth and tender in texture, rich in nutrition and soft and glutinous in taste.

▶ 此菜荣获全国烹饪大赛、国际烹饪大赛金奖。
It won the gold medal of the national and international cooking competitions.

宝参冬瓜盅

Braised Sea Cucumber and White Gourd

主料：冬瓜2500克、水发鱼肚200克、水发海参200克、水发竹荪150克、鲜贝200克、干贝150克。
配料：火腿50克、冬笋150克、鲜蘑200克、海米150克、草菇150克、人参150克。
调料：清汤1250克、精盐5克、料酒5克、胡椒粉2克、姜汁25克。

Main ingredients: 2500g white gourd, 200g fish maw, 200g sea cucumber, 150g bamboo shoots, 200g fresh scallops, 150g dried scallops.
Ingredients: 50g ham, 150g winter bamboo shoots, 200g fresh mushrooms, 150g dried shrimps, 150g button mushrooms, 150g ginseng.
Seasonings: 1250g broth, 5g refined salt, 5g cooking wine, 2g ground pepper, 25g ginger sauce.

制作方法 / Steps

1. 冬瓜洗净，雕刻成二龙戏珠图案，用开水烫一下待用。
2. 将水发海参、鱼肚片成抹刀片，竹荪切象眼片，火腿、冬笋、鲜蘑均切成丁，草菇一切为二，海米发好、干贝去筋蒸透，均用鸡汤煨透入味备用。
3. 将主、配料中的原料用清汤加调料煨透入味，装入冬瓜盅内，上屉蒸透即可食用。

1. Wash the white gourd and carve them into two dragons and a pearl on them, scald it with boiling water, set aside.
2. Slice the sea cucumber and fish maw, and slice bamboo fungus into parallelograms. Dice the ham, winter bamboo shoots and fresh mushrooms, halve the button mushroom, soak the dried shrimps, remove the tendon of dried scallops and steam the scallops thoroughly. Simmer them with chicken soup to tasty.
3. Put the simmered ingredients into the white gourd soup and steam them well. The dish is ready to be served.

特点：制作精细、造型美观、质地各异、汤鲜味美、滋补佳品。
Features: Fine production, delicate appearance, diversified texture, delicious soup, abundant nutrition.

红烧大拐枣裙边

Braised Turtle Rim with Calligonum Giganteum

主料：水发裙边750克、大拐枣75克。
配料：胡萝卜75克、莴笋75克。
调料：鸡汤750克、红花汁6克、糖3克、料酒5克、水淀粉25克、盐4克。

Main ingredients: 750g water-fat turtle rim and 75g calligonum giganteum.
Ingredients: 75g carrot and 75g asparagus lettuce.
Seasonings: 750g chicken soup, 6g safflower juice, 3g sugar, 5g cooking wine, 25g water starch and 4g salt.

制作方法
Steps

1. 将裙边改刀成长方块，胡萝卜去皮，切成门墩方丁，和去皮莴笋刻成八棱形空心球状，焯水入味。大拐枣洗净，用鸡汤蒸制入味备用。
2. 将裙边洗净，用鸡汤煨制入味备用。
3. 勺内放油烧热，加入主料和配料，放入红花汁、糖、料酒、盐、水淀粉勾芡，待成熟后取出，码入大盘中间，在两侧摆放煨透入味的胡萝卜门墩丁和八棱空心球，把大拐枣放在裙边上方，收浓芡汁，放在裙边上即可。

1. Cut the turtle rim into panes, peel carrots and cut them into squares, peel asparagus lettuce and cut them into octagonal hollow spheres; then blanch them to taste. Wash calligonum giganteum and steam it with clear soup to taste for later use.
2. Wash the turtle rim and simmer it in clear soup for later use.
3. Heat the oil in the spoon, add the main ingredients and ingredients, thicken the sauce with safflower juice, sugar, cooking wine, salt and water starch, take it out and put it into the middle of the large plate when it is well-cooked; place the stewed carrot squares and octagonal hollow balls on both sides, place the calligonum giganteum above the turtle rim, and put the thickened sauce on the turtle rim.

特点：汁浓味厚、裙边软烂浓香。
Features: The dish tastes richly flavoured with thickened sauce and the turtle rim tastes soft and tender.

▲ 此菜荣获全国烹饪大赛、国际烹饪大赛金奖。
It won the gold medal of the national and international cooking competitions.

翡翠三色鱼丸
Tri-colored Fish Balls

主料：活鱼肉750克。
配料：油菜心300克、枸杞15克、清汤750克、鸡蛋清125克、绿菜叶125克。
调料：番茄汁125克、葱姜油75克、白胡椒粉3克、精盐5克、黄酒5克、白糖3克、水淀粉15克。

Main ingredients: 750g live fish.
Ingredients: 300g oilseed rape, 15g lycium chinense, 750g broth, 125g egg white, 125g green vegetable leaves.
Seasonings: 125g tomato sauce, 75g scallion and ginger oil, 3g ground white pepper, 5g refined salt, 5g millet wine, 3g sugar, 15g water starch.

制作方法 / Steps

1. 先将鳜鱼肉开成两片，去脊骨刺，制成蓉状备用。
2. 将鱼蓉放入容器内，加入水、精盐、白糖、胡椒粉、葱姜油、水淀粉、蛋清，顺时针搅拌均匀，打上劲，分成3份备用。
3. 先将绿菜叶取汁，制成绿色鱼蓉，1份加入番茄汁制成红色鱼蓉，将3种颜色的鱼蓉氽成鱼丸，放入盘中。
4. 将油菜心焯至入味，点缀枸杞，码放在鱼丸周围，汤锅放入清汤调好味，加入精盐、黄酒、白糖、水淀粉勾芡，点上葱姜油，浇在鱼上丸即可。

1. Slice the mandarin fish into two halves, remove the bones and mince the fish, set aside.
2. Put the minced fish into a container, add water, refined salt, sugar, ground pepper, scallion and ginger oil, water starch and egg white. Mix them up clockwise, and divide the mixture into three portions, set aside.
3. Squeeze the green vegetable leaves for juice and color the minced fish in green, color another portion of minced fish in red with tomato juice. Shape the three portions of minced fish into balls and put them on a plate.
4. Fry the oilseed rape and garment it with lycium chinense and put them around the fish balls. Pour water into the pot to adjust the taste. Season it with refined salt, millet wine, sugar. Add water starch to thicken the sauce. Add scallion and ginger oil. Pour the sauce on the fish balls.

特点：鱼丸滑嫩富有弹性，入口即化，味道鲜美，色泽悦目。
Features: Fish balls are slippery, tender, elastic. They can melt in your mouth with delicious taste and pleasant colors.

▼ 此菜经北京烹饪协会认定，列为京贵八珍宝鼎宴名菜。
It is a famous dish of Beijing Cuisine Association and Beijing Eight Delicacies Banquet.

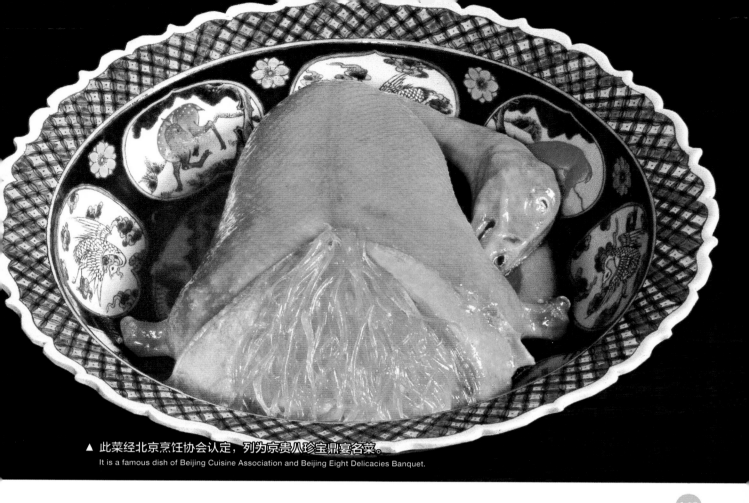

▲ 此菜经北京烹饪协会认定，列为京贵八珍宝鼎宴名菜。
It is a famous dish of Beijing Cuisine Association and Beijing Eight Delicacies Banquet.

北京填鸭包鱼翅
Shark Fin Wrapped in Peking Duck

主料：水发鱼翅2000克、脱骨北京填鸭2750克。
配料：母鸡250克、猪肘子250克、水发干贝150克、火腿100克、鲜龙须菜150克、冬笋150克。
调料：精盐6克、料酒10克、酱油15克、白糖10克、胡椒粉2克、葱15克、姜15克、鸡汤2500克。

Main ingredients: 2000g shark fin, 2750g boneless Peking duck.
Ingredients: 250g hen, 250g pork leg, 150g dried scallops, 100g ham, 150g fresh asparagus, 150g winter bamboo shoots.
Seasonings: 6g refined salt, 10g cooking wine, 15g soy sauce, 10g sugar, 2g pepper, 15g scallion, 15g ginger, 2500g chicken broth.

制作方法
Steps

1. 将水发鱼翅洗净并漂去腥味，用纱布包好备用。
2. 将猪肘子、母鸡斩成块状、用水焯透，放入锅中，加入鸡汤、火腿、干贝、葱、姜、酱油、精盐、料酒，将鱼翅包放在当中，用小火煨透入味。
3. 将脱骨鸭用八成热的水烫一下，并用水冲净，然后在鸭内外用盐搓上味，将煨好的鱼翅从脱骨鸭脖子开口处装入鸭内，用竹签别上口，鸭身处均匀抹上料酒，用八成热的油将鸭浇炸至成金黄色，备用。
4. 锅内加入鸡汤、精盐、料酒、胡椒粉、白糖，微火煮至酥烂，取出装入盘内。
5. 将龙须菜和冬笋焯透入味，码在鸭身下面两侧。原汤去沫燁浓，浇在鸭上即成。

1. Wash the shark fin clean, rinse to remove the fishy smell, wrap it up with gauze, and set aside.
2. Chop the pork knuckle and hen into chunks, blanch them, put them in a pot and add chicken broth, ham, scallops, scallion, ginger, soy sauce, refined salt, and cooking wine. Put the wrapped shark fin into the pot, and simmer with low heat.
3. Blanch the boneless duck with hot water, rinse it with water, and rub the inside and outside of the duck with salt. Put the simmered shark fin into the duck from the hole in the neck, seal the hole with bamboo sticks. Rub cooking wine evenly on the duck body, fry the duck with hot oil to golden brown, set aside.
4. Add chicken broth, refined salt, cooking wine, pepper, and sugar into the pot, boil to tender with slow fire, and then dish up.
5. Blanch the asparagus and winter bamboo shoots to tasty, and place them around the duck body. Skim the scum from the soup, simmer to thick, and finally pour the soup on the duck.

特点：鸭肉味美酥烂，鱼翅润滑鲜浓，系中华传统特色美味，滋补佳品。
Feature: the duck meat is delicious and soft, and the shark fin is smooth and tasty, which make it a delicious and nourishing traditional Chinese course.

▲ 此菜经北京烹饪协会认定,列为京贵八珍宝鼎宴名菜。
It is a famous dish of Beijing Cuisine Association and Beijing Eight Delicacies Banquet.

八珍龙锅

Soup of Eight Delicacies

主料： 鲍鱼250克、水发鱼翅400克、水发干贝300克、水发辽参400克、水发鱼肚300克、水发天然猴头蘑300克、水发羊肚菌200克，水发竹荪200克。
配料： 火腿25克、冬笋25克、鲜龙须菜25克、老鸡250克、排骨250克。
调料： 清汤1500克、奶汤400克、精盐4克、料酒15克、姜汁15克。

Main ingredients: 250g abalone, 400g shark fin, 300g scallop, 400g sea cucumber, 300g fish maw, 300g natural hericium mushroom, 200g morel mushroom, 200g bamboo fungus.
Ingredients: 25g ham, 25g winter bamboo shoots, 25g fresh asparagus, 250g chicken, 250 ribs.
Seasonings: 1500g broth, 400g creamy soup, 4g refined salt, 15g cooking wine, 15g ginger juice.

制作方法
Steps

1. 先将鲍鱼泡好，洗涮干净，用老鸡、排骨煲好备用。再将水发鱼翅、辽参、鱼肚、猴头蘑、羊肚菌、水发竹荪分别洗净，用清汤煨透入味备用。
2. 再将干贝去根洗净蒸透，要保持形状备用。
3. 再将龙须菜、火腿、冬笋分别改刀，焯透入味。
4. 将以上原料整齐地码入砂制龙锅内，调入清汤和奶汤、盐、料酒、姜汁，再加蒸干贝的原汤，双壁砂锅上火，小火煨至汁浓味厚时，即可上桌食用。

1. Soak the abalone first, rinse and clean it well, and boil it with old chicken and spare ribs, set aside. Wash the shark fin, sea cucumber, fish maw, hericium mushroom, morel mushroom, and bamboo fungus separately, and simmer thoroughly with chicken broth, set aside.
2. Remove the muscles of the scallops, wash and steam the scallops thoroughly while keeping their shape, set aside.
3. Shape the asparagus, ham, and winter bamboo shoots and blanch them to tasty separately.
4. Put the above ingredients neatly into a casserole, add broth and creamy soup, and add the soup of steamed scallop, simmer until the soup is thick, and then the course is ready to be served.

特点：八珍龙锅味浓香、质地各异、造型美观，具有极高的营养价值，是京贵八珍宝鼎名菜。
Feature: Soup of Eight Delicacies has strong flavor, exclusive mouthfeel, pleasing appearance, extremely rich nutritional value, and is a famous course in Beijing.

黄焖通天翅
Braised Shark Fin

主料：水发披刀翅1块1500克。
配料：鸡500克，鸭500克，肘子500克、干贝100克、火腿30克。
调料：清汤750克、料酒5克、精盐5克、水团粉15克、葱姜油15克、鸡油5克、葱50克、姜50克、酱油3克。

Main ingredient: 1 shark fin (1500g).
Ingredients: 500g chicken, 500g duck, 500g pork knuckle, 100g scallop, 30g ham.
Seasonings: 750g broth, 5g cooking wine, 5g refined salt, 15g water starch, 15g scallion and ginger oil, 5g chicken fat, 50g scallion, 50g ginger, and 3g soy sauce.

制作方法
Steps

1. 披刀翅洗净后加入葱姜，上屉蒸透，然后漂去腥味，控净水分，用纱布把鱼翅包好备用。
2. 鸡鸭肉斩成块状，放入锅中焯透控水。
3. 大砂锅内放入竹达，放入鸡、鸭肉、肘子、干贝，包好的鱼翅放中间，再加入清汤、葱、姜、料酒、盐、火腿，炖至鱼翅软烂入味。
4. 汤锅上火，放入清汤，鱼翅放入锅内，加入调味品，加入少量酱油，调好颜色，拢芡大翻锅，点入葱姜油，装盘即可。

1. Wash the shark fin, and put it on the tray with scallion and ginger, steam them thoroughly. Then rinse to remove the fishy smell, drain off the water, and then wrap the shark fin up with gauze, set aside.
2. Chop the chicken and duck into chunks, blanch them and drain off the water.
3. Put a bamboo grid in a large casserole, add chicken, duck, pork knuckle, and scallops, and put the wrapped shark fin in the middle. Then add chicken broth, scallion, ginger, cooking wine, salt, ham, and stew until the shark fin become soft and tasty.
4. Put the pot on the heat, add broth and shark fin into the pot, add the condiments, add a small amount of soy sauce, adjust the color, turn the pot, thicken the sauce, and add a little scallion and ginger oil.

特点：色泽软烂入味、不失其形、翅香滑润、汁浓味厚。
Feature: The dish is soft and tasty without losing their shapes, the fin is flavorful and tender, and the sauce has a strong flavor.

▼ 此菜经北京烹饪协会认定、列为京贵八珍宝鼎宴名菜。
It is a famous dish of Beijing Cuisine Association and Beijing Eight Delicacies Banquet.

▲ 此菜经北京烹饪协会认定，列为京贵八珍宝鼎宴名菜。
It is a famous dish of Beijing Cuisine Association and Beijing Eight Delicacies Banquet.

龙须面
Longxu Noodle

主料：面粉1500克、南瓜1个1500克。
调料：食用油150克，青菜汁15克，矿泉水750克（分2次使用）、精盐4克，白糖15克。

Main ingredients: 1500g flour and 1 pumpkin (1500g).
Seasonings: 150g cooking oil, 15g vegetable juice, 750g mineral water (use it in 2 times), 4g refined salt, 15g sugar.

制作方法
Steps

1. 盆内放入面粉，加水、糖、盐揉成面团，不能窝水，醒发后，分成三份，制成三种不同颜色的面团备用，把南瓜雕刻成龙备用。
2. 三种不同颜色的面团溜面时，用力要均匀，抻好面上条要稳而准，抻面时不能去条、粘连、粗细不均。
3. 锅内放少量油，先将两种龙须饼炸好后控油。
4. 然后再用大锅放油，把整把面炸好装盘并码在大盘中间，两侧码放不同颜色的龙须饼，周围点缀雕刻而成的龙和法香即可。

1. Put flour in the basin, add water (don't add too much), sugar, and salt to knead dough. After the dough is proofed, divide it into three parts to make three different colors, set aside. Carve the pumpkin into a dragon shape, set aside.
2. When pulling the noodles in three different colors, the force should be even, the technique should be stable and accurate, and the noodles should not be stripped, sticky, or uneven in thickness.
3. Put a small amount of oil in the pot, and fry two sets of Longxu Noodles into breads and then drain off the oil.
4. Then add oil in a big pot, fry a bundle of noodles and put it in the middle of the big plate. Place the two breads of different colors on both sides, and decorate with the pumpkin dragon and parsley.

特点：龙须面香脆可口，整体美观大方。
Feature: Longxu Noodle is crispy and flavorful, and looks pleasing and delicate.

海参烧鹿筋
Stewed Sea Cucumber and Deer's Sinew

主料：辽参750克、刺参500克、水发鹿筋500克。
配料：青蒜段15克。
调料：盐3克、味精2克、酱油15克、料酒10克、糖3克、鸡汤400克、水团粉35克、浓香油10克。

Main ingredients: 750g sea cucumber, 500g stichopus japonicus, 500g deer's sinew.
Ingredients: 15g green garlic.
Seasonings: 3g salt, 2g monosodium glutamate, 15g soy sauce, 10g cooking wine, 3g sugar, 400g chicken broth, 35g water starch, 10g sesame oil.

制作方法
Steps

1. 将水发海参、水发鹿筋洗净后，海参改刀1分为2，蹄筋切1寸5分的段，用鸡汤煨透入味，备用。
2. 锅内重新放入鸡汤，加入以上调味品用小火煨至软烂入味，并加入浓香油。
3. 煨透的海参、鹿筋拢芡，加入青蒜段，投装入盘中即可。

1. After washing the sea cucumbers and deer's sinews, cut the sea cucumber into two pieces, and cut the sinews every 1.5 inch, simmer them in chicken broth, set aside.
2. Add another portion of chicken broth in the pot, add all the condiments and simmer over low heat until the ingredients become soft and flavorful, and add sesame oil.
3. Thicken the simmered sea cucumbers and deer's sinews, add green garlic, and then dish up.

特点：山珍和海味同烧，味道各异，软烂浓香，色泽美观。
Feature: Stewed delicacies from land and water, the flavor is exclusive, the mouthfeel is soft and smooth, and the color looks pleasing.

▼ **此菜经北京烹饪协会认定，列为京贵八珍宝鼎宴名菜。**
It is a famous dish of Beijing Cuisine Association and Beijing Eight Delicacies Banquet.

▼ 此菜经北京烹饪协会认定，列为京贵八珍宝鼎宴名菜。
It is a famous dish of Beijing Cuisine Association and Beijing Eight Delicacies Banquet.

游龙戏凤

Fried Lobster and Large Prawns and Chicken

主料：龙虾1只2500克、大对虾12只。
配料：鸡里脊500克、鸡蛋150克、油菜心150克。
调料：葱姜25克，面粉25克，水团粉125克，清汤250克，姜汁10克，胡椒粉2克，葱姜油15克，味精2克、糖2克，食用油适量。

Main ingredients: 1 lobster (2500g), 12 large prawns.
Ingredients: 500g chicken tenderloin, 150g egg, 150g oilseed rape.
Seasonings: 25g scallion and ginger, 25g flour, 125g water starch, 250g broth, 10g ginger juice, 2g pepper, 15g scallion and ginger oil, 2g monosodium glutamate, 2g sugar and a proper amount of edible oil.

制作方法
Steps

1. 先将龙虾宰杀好，将肉取下，改成块状，用蛋清、水团粉浆好，再将大虾中间去皮，开一刀，去沙包、沙线，翻成龙脱袍形，洗净入味备用。
2. 将鸡里脊改刀成条状，加入鸡蛋、水团粉、面粉抓好糊后，炸好，放入大盘中间，再将龙虾头部用葱姜余透入味，控去水分，放在大盘两端。
3. 再将12只大虾上屉蒸好，码在大盘两边，再把龙虾肉过油滑透，加调味品爆炒入味，放在大盘中间的鸡柳上，两边放油菜心，浇汁即可。

1. 1. First slaughter the lobster, pick out the meat, and cut it into lumps and wrap up with egg white and water starch. Then cut the prawns from the middle of the back, remove sand veins, and wash them well and set aside.
2. Cut the chicken tenderloin into strips, wrap them with egg and starch, fry them and put in the middle of the plate. Blanch the lobster's head to tasty, drain off the water, and place it on the side of the plate.
3. Put 12 prawns on a tray and steam them, and place them on both sides of the plate; stir-fry the lobster meat to tasty, put it on the chicken tenderloin strips in the middle of the plate, add the oilseed rape on both sides, and add sauce.

特点：白绿相间，龙虾在上为龙、鸡在下面为凤，造型美观大方，是高档宴会八珍宝鼎宴名菜之一。
Feature: The color is mixed with white and green; the lobster represents the dragon, and the chicken represents the phoenix; the course has a delicate look, and is one of the famous courses in luxury feasts.

年表

Chronology

1952年8月27日，出生在北京市大兴县亦庄。

1960年至1965年，在亦庄小学读书，小学毕业。

1966年至1968年，在北京市大兴县鹿圈中学读书，中学毕业。

1969年1月23日，在北京市大兴县亦庄大队双桥插队。

1972年7月，插队返城，被分配到北京市宣武区饮食服务公司所属的华北楼党支部合算店，成为一名学徒。后来被调到瑞宾饭庄工作（原名祥瑞饭馆）。

1978年3月，被调往前门西大街2号楼东侧正阳春任副经理兼厨师长。

1979年，参加北京饮食公司首批山东菜系厨师进修班。

1980年11月18日，任正阳春经理兼厨师长。

1979年至1982年，在正阳春任厨师长、副经理期间，多次为戏曲名家李万春、李小春父子烹制高级宴会和便餐。

1980年至1982年，在正阳春任厨师长、经理期间，多次为戏曲艺术家魏喜奎、周桓伉俪烹制宴会和便餐。

1983年2月，开始筹备恢复泰丰楼。

1983年3月，找师父艾长荣恢复瑞安楼。艾长荣同意在门框胡同9号注册成立。

1984年8月，在北京市宣武区职工业务技术选拔赛中荣获北方炒菜项目第一名。

1984年3月8日，泰丰楼正式恢复开业。李启贵任厨师长。

1984年5月14日，经宣武科协鉴定考核，颁发烹饪技术优秀证书。

1984年，对中国烹饪颇有研究的爱新觉罗·溥杰先生在泰丰楼用完餐后称"饭菜味道堪称上等"，并以"继往开来，发扬光大"八字书法条幅相赠。

1984年11月，参加了中国烹饪界第一部电影《吃与文化》拍摄。

1984年5月16日，任泰丰楼厨师长，为前来考察的全国人大常委会副委员长朱学范烹制宴会。朱学范当场题词："色香味俱佳，服务上乘"。

1985年，经北京市劳动局考核，名列第一，被北京市破格提拔为特二级技师。

1986年，李启贵大师在北京军区六分部第二期厨师培训班讲课。

1986年1月，荣获奥林匹克第五届世界烹饪大赛金牌、证书及纪念杯。

1986年2月28日，世界烹饪协会秘书长在卢森堡大使馆会见中国代表团。

1986年，在卢森堡参加奥林匹克第五届世界烹饪大赛，荣获金牌后和世界大赛秘书长合影留念。

1986年2月，参加奥林匹克第五届世界烹饪大赛，获得金牌后应邀到卢森堡帝国饭店表演中国的烹饪技艺。

1986年2月，在卢森堡国际俱乐部为27个国家的厨师表演传授中国的烹饪技艺。

1986年3月，在卢森堡大使馆为卢森堡首相烹制了首相宴。

1986年3月，在卢森堡北京楼献艺烹制北京菜，引起轰动。

1986年3月，荣获北京市人民政府财贸办公室表彰。

1986年9月22日，荣获北京烹饪协会颁发的营养培训班证书。

1987年4月，应南郊农场邀请，举办了厨师培训班，培养厨师队伍，受到热烈欢迎。

1987年1月，获得北京市饮食服务总公司技师联谊会表彰。

1987年10月，在宣武区培训中心开展了北京市直观教学。

1987年12月，被评为宣武区成人教育先进工作者。

1988年9月10日，荣获北京市饮食服务总公司评选的1987年度职工教育先进工作者。

1988年，荣获北京首届烹饪技术比赛展台表演奖。

1988年11月，参与编写了《北京百店千家菜》一书。

1988年12月21日，荣获北京首届烹饪技术比赛美食杯奖。

1989年12月，北京市饮食服务总公司特聘李启贵为高级业务技术培训烹饪技术讲师。

1990年至1992年，任职日本大阪南海假日酒店行政总厨。

1990年3月19日，获得北京市西城区日语专科学校成绩合格结业证书。

1990年，在日本大阪电视台和日本著名主持人渡边撒主持中华厨艺节目，现场烹制清汤官燕、酥炸蝎子、雪花龙须面。

1990年10月29日，在日本大阪接待北京市饮食服务总公司代表团。

1990年11月14日，在日本大阪推出了北京厨师长10款京菜美食节。

1991年4月，在日本白滨湾点心工厂，表演芸豆卷、豌豆黄、龙须面、佛手酥等中国宫廷名点的制作。

1991年6月8日，在大阪南海假日酒店为五星级酒店总裁、总厨及社会各界名人表演烹制中国名菜。

1991年9月，在日本大阪推出了甲鱼全宴美食活动。

1991年12月1日至1992年1月31日，参与两款高级宴会：福喜、寿喜高级宴会美食节活动。

1992年，加入中国共产党。

1992年4月，在日本大阪为接待日本著名相扑运动员烹制特级宴会。

1992年5月29日，荣获日方颁发的特级厨师长证书。

1992年6月16日，参与扶贫单位门头沟饮食公司燕乡楼开业仪式。

1993年2月23日，被北京市饮食服务总公司聘为培训中心专业教师。

1993年2月27日，被新疆吐鲁番绿洲宾馆聘为讲师。

1993年4月，在北京市宣武区争办奥运烹饪技术大赛中荣获热菜一等奖。

1993年8月，被北京市聘为北京争办奥运会烹饪技术大赛评委。

1993年8月，荣获北京争办奥运会烹饪技术大赛纪念奖。

1993年9月2日，荣获中国烹饪协会举办的京、沪、川、粤烹饪技能比赛第一名。

1993年11月1日，在第三届全国烹饪大赛中荣获个人热菜比赛金牌及证书。

1993年12月，在1993年度宣武区开展的职工岗位练兵技术比赛活动中，取得汉民炒菜第一名，荣获技术能手称号。

1993年12月，在宣武区开展的献计策、增效益、创三优立功竞赛活动中，被评为区级标兵。

1993年12月10日，在第三届全国烹饪技术比赛团体赛中荣获银牌。被中国烹饪协会授予全国优秀厨师称号。

1994年1月，被北京翔达饮食（集团）公司评选为1993年劳动竞赛标兵。

1994年1月28日，被北京商学院聘为北商一分部客座教授。

1994年1月，被北京市总工会评为技术能手，授予爱国立功标兵称号。

1994年2月，被宣武区科学技术协会评为1993年度科协互作积极分子。

1994年6月，参与出版了台湾编写的《中华名厨招牌菜（第二）北京篇》。

1994年10月1日，荣获北京市劳动局颁发的高级厨师任职资格证书。

1994年至1995年，任泰丰楼饭庄副经理兼厨师长。

1995年2月6日，获得中餐烹调技师资格证书。

1995年3月，被北京市宣武区科学技术协会评为1994年度宣武区科协系统先进个人。

1995年3月，被北京翔达饮食（集团）公司评选为1994年度优秀共产党员。

1995年5月，获得中华人民共和国劳动部颁发的中餐烹调高级技师证书。

1995年5月25日，被聘为中直机关工人考核委员会烹饪专业考评委员。

1995年7月，被北京市宣武区人民政府授予宣武区第二届有突出贡献的科教、技术、管理人才称号。

1995年，在泰丰楼副经理岗位上挂职到日本东京美味斋任厨师长。1996年回国。

1996年2月26日，被评为北京翔达饮食（集团）公司1995年度优秀共产党员。

1996年12月，被北京市爱国卫生运动委员会评为北京市爱国卫生先进工作者。

1997年3月，被评为1996年度宣武区行业状元。

1997年7月17日，被聘为庆祝香港回归北京名菜名点鉴定委员会委员。

1997年，任北京天伦王朝饭店技术顾问。

1997年12月26日，加入京华名厨联谊会。

1998年1月，被北京市爱国卫生运动委员会评为北京市爱国卫生先进工作者。

1998年，参加国家内贸部召开的起草中国烹饪大师的标准制定工作。

1998年5月13日，被聘为北京市饮食业工人技师考评委员会评委。

1998年8月31日，被聘为房山饭店技术顾问。

1998年9月，被聘为北京市流水养鱼大奖赛总裁判长。

1998年9月，荣获北京市劳动局颁发的职业技能鉴定考评员证书。

1998年11月24日，率团到香港京华国际酒店献艺表演。香港23家新闻媒体争相报道，轰动香港。

1999年3月，在天伦王朝饭店接待了加拿大、中国香港特别行政区和中国大陆三方十余人的股东的重要而特殊的宴会。

1999年4月6日，被聘为北京第三届烹饪服务技术大赛评委。

1999年6月30日，获得中共中央党校函授学院大专班毕业证书。

1999年6月30日，加入北京烹饪协会。

1999年9月，被国家国内贸易局认定为烹调专业国家一级评委。

1999年9月，被北京市商业委员会聘为"与祖国同庆，北京首届燕京啤酒杯大众美食节评委"。

1999年9月29日，亲自主理北京市授予霍英东、李嘉诚"北京市荣誉市民"的500人大型宴会。

1999年10月，在有38家五星级酒店参加的天伦王朝酒店管理集团首届烹饪大奖赛上，任总裁判长。

1999年11月12日，被聘为第四届全国烹饪大赛个人赛热菜评委。

1999年，天伦王朝酒店管理集团总裁为李启贵颁发纯金五星奖章。

2000年2月13日，出席由侯耀文主持的中央电视台"中国大陆、香港、澳门、台湾四位烹饪大师施绝技贺新春节目"录制活动。李启贵大师现场烹制九龙八珍宝鼎。

2000年，在电视台和著名演员谢园、李琦、尹相杰共庆新春佳节活动。

2000年，应邀出席挪威大使馆贺新春活动，献艺操作"风生水起"。

2000年，应邀到上海锦江饭店为挪威首相烹制首相宴。

2000年3月，在日本东京，和世界烹饪联合会会长张世尧、杨柳会长、林则普秘书长出席颁奖现场，和本届大赛个人赛、团体赛两个世界金牌获得者合影。

2000年3月6日，参加在日本东京举行的第三届中国烹饪世界大赛，任北京所有的3个代表队（包括北京全聚德、北京功德福、北京大董烤鸭店）的总领队。

2000年3月9日，被日本东京组委会聘为第三届中国烹饪世界大赛评判委员会委员。

2000年3月30日，获得国家国内贸易局授予的中国烹饪大师称号。

2000年6月，被聘为北京旅游集团青年技能大赛评委。

2000年7月，中国驻美大使柴泽民先生为李启贵大师题词："树天伦品牌，创一流佳绩"。

2000年8月1日，被北京烹饪协会聘为北京名菜名点鉴定师。

2000年8月8日，被北京东方美厨饮食文化交流中心聘为常务理事。

2000年9月22日，北京市商委领导与饮服业国家级评委、大师名师合影。

2000年10月，在北京电视台和曹为、英壮、美国钟生做评委。

2000年10月，被北京功德福餐饮管理有限公司聘为技术顾问，并和全国著名演员共庆北京龙乡情大酒店开业。

2000年10月26日，应邀前往德国交流考察德国精品酒店无人服务流程。

2000年11月2日，被首届中国美食节组委会评定为2000年中国十佳烹饪大师。

2000年11月7日，参加中国烹饪协会餐饮考察团，前往意大利考察。

2000年11月23日，在香港加入法国蓝带美食会。

2000年11月，在北京天伦王朝饭店接待前来检查工作的全国政协副主席孙孚凌，并介绍了专利菜肴及经营特色。

2000年，入选中国餐饮服务年鉴。

2000年12月24日，在北京天伦王朝饭店亚太第一室内广场成功扮演圣诞国王，带领"公主"庆贺圣诞节。

2001年，在天伦王朝饭店接待来访合作的法国、毛里求斯、墨西哥、乌拉圭四国大使和法国驻远东主席古扎礼先生。

2001年8月20日，荣获中国旅游饭店协会颁发的全国饭店餐饮经营管理高级研修班证书。

2001年10月12日，加入中国烹饪协会名厨专业委员会（中国名厨联谊会）。

2001年10月，参加首届中国烹饪大师名师精品集的编写。

2001年10月，应邀前往比利时交流考察中西合璧特色菜肴。

2001年10月，在《京华名厨论文集》中，发表了继承弘扬北京老字号的烹饪文化特色论文。

2001年10月，入选首届中国烹饪大师名师精品集。

2001年11月8日，举行首届挪威三文鱼面点大奖赛，被聘为总裁判长。

2002年6月26日，被世界中国烹饪联合会认定为中餐烹饪技术比赛国际评委。

2002年，法国驻远东主席古扎礼和中国驻法国大使授予中国十佳烹饪大师李启贵法国蓝带奖。

2002年6月，在第四届中国烹饪世界大赛上，张世尧会长为优秀国际评委李启贵颁发奖杯。

2002年10月，北京天伦王朝饭店推出轰动全国的"六大名厨闹京都美食节"活动。

2002年6月26日，被世界中国烹饪联合会聘为第四届中国烹饪世界大赛国际评委，该大赛在马来西亚吉隆坡举行。

2002年10月8日，世界烹饪联合会、中国烹饪协会负责人和李启贵大师参加了对天伦王朝精品满汉全席的认证考评。

2002年，原商业部副部长姜习为李启贵大师题词："集各家之长，创启贵之佳肴"。

2003年，在中国人民银行总行，亲自主理由国务院总理温家宝和世界银行行长出席的500人大型会议餐，受到一致好评。

2003年3月8日、9日，第三届中国烹饪世界大赛在日本东京举行，李启贵任热菜组国际评委。

2003年5月，获得第五届全国烹饪技术比赛的执裁资格。

2003年7月，世界烹饪协会会长兼中国烹饪协会会长为李启贵大师题词："为弘扬中华饮食文化做出积极贡献"。

2003年9月3日，应邀到满洲里收徒刘忠新，并献艺金秋美食节。

2003年9月，被北京民族大学聘为兼职教授。

2004年4月10日，被聘为第五届全国烹饪技术比赛个人赛总决赛评委。

2004年8月，在北京天伦王朝饭店接待了前来访问的西班牙大使。

2004年9月18日，在国际中餐名厨创新菜展示活动中制作的蜂蓉凤尾虾荣获铂金奖。

2004年9月，主编了《中华名菜精选》一书。

2004年11月18日，被世界中国烹饪联合会聘为第五届中国烹饪世界大赛国际雕刻组评委。

2005年，入选名厨专业委员会、中国烹饪协会出版的《华夏全宴》一书。

2005年，出席在人民大会堂举行的第五届全国高新技术比赛颁奖大会。

2005年6月8日，参与北京市工贸技师学院烹饪技师教学方案的制定。

2005年，法国厨皇协会授予李启贵大师五星白金奖。

2005年10月18日，被中国烹饪协会认定为全国餐饮业中式烹调一级评委。

2005年11月，李启贵大师荣获法国厨皇协会金奖。

2006年2月，参加在河北省举办的中国餐饮服务大师、名师认定，任认定专家组成员。

2006年3月，被北京烹饪协会授予北京特级烹饪大师称号。

2006年11月28日，主持召开天伦集团第四届烹饪大赛总裁判长。

2006年12月9日，参加国家职业技能竞赛裁判员评委培训班。

2006年12月，被北京烹饪协会认定为北京餐饮特技评委。

2007年6月17日，获得中华人民共和国劳动和社会保障部职业技能鉴定中心职业技能竞赛裁判员资格。

2007年6月26日，被中国烹饪协会认定为全国餐饮业一级评委。

2007年10月，在荷兰鹿特丹海上乐园参加了华人华侨庆祝国庆大型活动。

2007年10月19日，应邀前往荷兰的格罗宁根市献艺，该市的楼外楼中餐厅开业，中国驻荷兰领事、格罗宁根市的市长出席。

2007年11月27日，主持天伦王朝国际酒店管理集团第二届大奖赛，任总裁判长。

2007年，李启贵大师应邀到欧洲表演中华鼎宴，和美食家交流中西饮食文化、烹饪技艺。

2008年10月21日，北京燃气集团50年大庆，主理146桌大型宴会。

2008年8月9日，喜迎百年奥运，为德国前总理施罗德烹制中华鼎宴。

2008年8月12日，荣获挪威海产局颁发的特别贡献奖。

2008年10月，被世界烹饪联合会评为"国际中餐大师"。

2008年6月，在北京烹饪协会第四届第五次理事会上，被增补为理事。

2008年9月16日，主理的天伦王朝"满汉全席"入选中国烹饪协会专家工作委员会编写的《中国菜单赏析》。

2008年10月16日，被世界烹饪联合会聘为第六届中国烹饪世界大赛国际评委，担任冷菜组组长。

2009年4月，被中国烹饪协会聘为第六届全国烹饪大赛评委。

2009年10月20日，加入世界烹饪联合会名厨专业委员会。

2009年10月，当选为世界中国烹饪联合会专业委员会第一届委员。

2009年8月27日，荣获台湾饮食文化"研习证书"。

2010年，参与编写了《中国大厨在海外》一书。

2010年，研发了中华八珍宝鼎，获得实用新型专利证书。

2010年4月20日，研究发明灵芝烤鸭，获得专利。

2010年4月，到中央军委培训厨师，讲烹饪，讲平衡膳食，讲营养，并授课献艺。

2010年2月4日，175个国家大使迎新春，在各国大使招待会上烹制了中华八珍宝鼎宴。

2010年6月，编著了《中国烹饪大师作品精粹专辑》。

2011年8月26日，举行了专利技术传承仪式暨60大寿，出席者有张世尧、苏秋成、杨登彦。

2012年，在万寿寺为澳大利亚前总理霍克烹制中华八珍宝鼎专利名菜。

2012年2月13日，在美国纽约成功表演了雪花龙须面、中华八珍宝鼎，美国前劳工部长莫天成、中国驻美国领事馆参赞徐兵、美国旅游局长罗伯特及佛罗里达州政府官员及来自各州的200多人出席。

2012年4月，荣膺中国烹饪协会中国烹饪大师金爵奖。

2012年4月17日，被新加坡聘为国际中餐筵席争霸赛评委会副主席。

2012年，荣获第七届中国烹饪世界大赛团体赛金牌。

2012年8月,被世界中国烹饪联合会聘为世界中国烹饪联合会国际中华厨艺高级研修班客座讲师。

2012年,在天伦国际酒店管理集团内蒙古鄂尔多斯伊泰凯瑞博国际酒店,荣获第七届中国烹饪世界大赛团体金奖。

2012年12月6日,恢复了北京著名八大楼之一的鸿庆楼饭庄。

2012年,被世界中国烹饪联合会聘为国际评委。

2012年9月,完成世界烹饪联合会第六期国际评委培训班,荣获评委证书。

2012年11月16日,被聘为第七届中国烹饪世界大赛国际团体赛评委。

2012年,荣获第七届中国烹饪世界大赛金牌评委奖。

2013年,任美丽中国创新菜总裁判长。

2013年8月28日,被天伦国际酒店管理集团聘为厨房运营总顾问。

2013年9月,在中国烹饪协会举办的捷克、斯洛伐克美食节上献艺表演。

2013年11月11日,被中国烹饪协会聘为第七届全国烹饪大赛昆明赛区热菜评审组组长,该区为全国最大的赛区。

2013年,任中国饭店业名师名菜大赛评委。

2013年,被聘为中国饭店采购供应协会餐饮专业委员会顾问。

2013年9月，在北京举行了环京马拉松拉力赛中国赛区的大型国际活动，被聘为龙泉宾馆顾问。

2014年，被聘为鲜锋好味道精品美食专家评委。

2014年，在麦德龙杯争霸赛上，荣获优秀裁判员奖。

2013年至2018年，为中南海服务处指导、培训、考评厨师队伍。

2014年12月17日，荣获中国食文化研究会授予的"餐饮文化功勋奖"。

2015年1月31日，荣获由8家协会评选的2014年度传播饮食文化特殊贡献奖。

2015年8月7日，被中国专业人才考评专家委员会注册为中国专业人才库高级酒店管理师。

2015年9月，被中国烹饪协会认定为资深级注册中国烹饪大师。

2016年5月，对中央组织部高级厨师进行了升级考评。

2016年6月，入选世界中餐业联合会成立25周年纪念专辑一书。

2016年6月，被世界中餐业联合会授予烽火杯中餐厨师艺术家称号。

2017年3月26日，中国烹饪协会会长姜俊贤主持发扬工匠精神李启贵大师技艺传承收徒仪式。

2017年5月，被中国烹饪协会评选为中国餐饮30年功勋人物。

2017年8月3日，参加中国烹饪协会专家组，考评渭南市、洛阳黄河生态美食名城，任组长。

2018年4月9日，被中国烹饪协会聘为考评大连市138位注册中国烹饪大师评委。

2018年5月，被中国烹饪协会授予改革开放40年中国餐饮行业技艺传承突出贡献人物。

2018年10月，经中国烹饪大师名人堂师徒传承工作指导委员会核准，资深级注册中国烹饪大师李启贵评为中国烹饪大师名人堂导师。

2018年10月20日，当选第六届中国烹饪世界大赛冷菜拼盘组评委组长。

2018年12月25日，中国烹饪协会姜俊贤会长为中华八珍宝鼎宴颁发证书，并题词："京贵八珍宝鼎宴"。

2018年11月5日，荣获中国烹饪协会职业技能竞赛注册裁判员A级证书。

2018年11月22日，被聘为中国烹饪协会考评吉林省长春市75位注册大师的评委。

2019年7月2日，被聘为中国烹饪协会举办的第八届全国烹饪大赛北京赛区的热菜评判组组长。

2019年9月24日，被聘为中国烹饪协会在天津举办的全国大赛总决赛评判组组长。

On August 27, 1952, he was born in Yizhuang, Daxing County, Beijing.

From 1960 to 1965, he studied in Yizhuang Primary School with graduation from the primary school.

From 1966 to 1968, he studied in Luquan Middle School in Daxing County, Beijing with graduation from the middle school.

On January 23, 1969, he went to live and work in Shuangqiao Village, Yizhuang Brigade, Daxing County, Beijing.

In July 1972, he returned to the city from the countryside and was assigned to the North China Building Party Branch Hotel under Xuanwu District Catering Service Company in Beijing, and became an apprentice. Later, he was transferred to Ruibin Restaurant (formerly known as Xiangrui Restaurant).

In March 1978, he was transferred to Zhengyangchun Restaurant on the east side of Building 2, Qianmen West Street as deputy manager and chef.

In 1979, he took part in the first batch of Shandong Cuisine Chef Training Classes by Beijing Catering Company.

On November 18, 1980, he served as Manager and Chef of Zhengyangchun Restaurant.

During his tenure as chef and deputy manager of Zhengyangchun Restaurant from 1979 to 1982, he cooked many high-grade banquets and light meals for famous opera artists Li Wanchun and Li Xiaochun, father and son respectively.

During his tenure as chef and manager of Zhengyangchun Restaurant from 1980 to 1982, he cooked banquets and light meals for opera artists Wei Xikui and Zhou Huan many times, who were couple.

In February 1983, he started to make preparations for the restoration of Taifeng Restaurant.

In March 1983, he asked Master Ai Changrong to restore Rui'an Restaurant. Ai Changrong agreed to register it at No. 9 Menkuang Lane.

In August 1984, he won the first place in the Northern Cooking Program during the Staff Business Technology Selection Competition in Xuanwu District of Beijing.

On March 8, 1984, Taifeng Restaurant officially resumed operation. Li Qigui assumed the chef.

On May 14, 1984, he obtained the Certificate of Excellence in Cooking Technology issued by Xuanwu Association for Science and Technology.

In 1984, Mr. Aisin Giorro Pu Jie, who had done a lot of research on Chinese cuisine, said after having the meal in Taifeng Restaurant that "the taste of dishes is excellent" and presented it with a calligraphy banner of "carrying forward the cause pioneered by predecessors and forge ahead into the future".

In November 1984, he took part in the filming of the first Chinese cooking movie *Food and Culture*.

On May 16, 1984, he served as Chef of Taifeng Restaurant and cooked a banquet for Zhu Xuefan, Vice Chairman of the Standing Committee of the National People's Congress, who came to inspect. Zhu Xuefan wrote the inscription on the spot, "Excellent Color, Aroma and Taste, and High-Quality Service".

In 1985, he was ranked first by the Beijing Municipal Labor Bureau (currently Beijing Municipal Bureau of Human Resources and Social Security) and was exceptionally promoted to be Special Second-Class Technician.

In 1986, he gave a lecture in the Chef Training Class Phase II of the 6th Division of the Beijing Military Region.

In January 1986, he won the gold medal, certificate and Memorial Cup in the 5th IKA.

On February 28, 1986, the Secretary-General of the World Cuisine Association met with the head of the Chinese delegation at the Embassy of Luxembourg.

After winning the gold medal in the 5th IKA in Luxembourg in 1986, he took a photo with the Secretary-General of the Competition.

In February 1986, he was invited to perform Chinese cooking skills at the Hotel Empire in Luxembourg, after winning the gold medal in the 5th IKA.

In February 1986, he taught Chinese cooking skills to chefs from 27 countries at Luxembourg International Club.

In March 1986, he cooked the Prime Minister's dinner for the Prime Minister of Luxembourg at the Luxembourg Embassy.

In March 1986, he performed in Beijing Building at Luxembourg to cook Beijing cuisine, causing a sensation.

In March 1986, he was commended by the Finance and Trade Office of the Beijing Municipal People's Government.

On September 22, 1986, he was awarded the Certificate of Nutrition Training Course issued by Beijing Cuisine Association.

In April 1987, at the invitation of Beijing Nanjiao Farm Co., Ltd., he gave lectures at the chef training class to train chefs, and was warmly welcomed.

In January 1987, he was commended by the Technician Association of Beijing Catering Service Corporation.

In October 1987, he demonstrated cooking skills in Xuanwu District Training Center.

In December 1987, he was named an advanced worker in adult education in Xuanwu District.

On September 10, 1988, he was awarded the 1987 Advanced Worker in Staff Education by Beijing Catering Service Corporation.

In 1988, he won the Booth Performance Award of the First Beijing Cooking Technology Competition.

In November 1988, he participated in the compilation of the book Beijing Hundreds of *Restaurants and Thousands of Cuisines*.

On December 21, 1988, he won the Gourmet Cup Award in the First Beijing Cooking Technology Competition.

In December 1989, he was specially employed by Beijing Catering Service Corporation as a lecturer in cooking technology for senior business technology training.

From 1990 to 1992, he served as Executive Chef of Holiday Inn Nankai, Osaka, Japan.

On March 19, 1990, he won the Certificate of Qualification of Japanese Junior College in Xicheng District, Beijing.

In 1990, he cooked Best-Quality Swallow's Nests in Clear Soup, Fried Scorpions and Snowflake Longxu Noodles at the scene of Chinese Cooking Program hosted by Japanese host Watanabe on Japan's Television Osaka.

On October 29, 1990, he cooked for a delegation from Beijing Catering Service Corporation in Osaka.

On November 14, 1990, he cooked at the Beijing Chefs' 10 Beijing Cuisine Festival in Osaka.

In April 1991, he performed the making of Kidney Bean Rolls, Pea Cake, Longxu Noodles, Crispy Finger Citron and other famous Chinese court cakes in Japan's Shirahama Bay Dim Sum Factory.

On June 8, 1991, he performed cooking famous Chinese dishes for the presidents and chefs of the five-star hotels and celebrities from all walks of life in Holiday Inn Nankai, Osaka, Japan.

In September 1991, he cooked a turtle banquet in Osaka.

From December 1, 1991 to January 31, 1992, he cooked at two high-level banquets: Fuxi and Shouxi high-level banquets and food festivals.

In 1992, he joined the Chinese Communist Party.

In April 1992, he cooked a special banquet in Osaka, Japan to receive famous Japanese sumo wrestlers.

On May 29, 1992, he was awarded the Certificate of Super Chef issued by Japan.

On June 16, 1992, he cooked for the opening ceremony of Yanxiang Restaurant of Mentougou Catering Company, a poverty alleviation unit.

On February 23, 1993, he was employed by Beijing Catering Service Corporation as a professional teacher in the training center.

On February 27, 1993, he was employed as a lecturer by Greenland Hotel in Turpan, Xinjiang.

In April 1993, he won the first prize for hot dishes in the Beijing Olympic Cooking Technology Competition in Xuanwu District of Beijing.

In August 1993, he was employed by Beijing Municipality as a judge of the Beijing Olympic Cooking Technology Competition.

In August 1993, he won the Memorial Award of Beijing Olympic Cooking Technology Competition.

On September 2, 1993, he won the first place in the Beijing, Shanghai, Sichuan and Guangdong Cooking Skills Competition held by the China Cuisine Association.

On November 1, 1993, he won the gold medal and certificate in the individual hot dish group in the 3rd National Cooking Competition.

In December 1993, he won the first place in cooking for the Han people and the title of technical expert in the Technical Competition for Staff & Workers' Post Training held in Xuanwu District in 1993.

In December 1993, he was rated as a district-level pacesetter in the Competition of Offering Suggestions, Increasing Benefits & Creating Three Excellent Meritorious Deeds in Xuanwu District.

On December 10, 1993, he won the silver medal in the team competition of the 3rd National Cooking Technology Competition. He was awarded the title of National Excellent Chef by the China Cuisine Association.

In January 1994, he was rated by Beijing Xiangda Catering (Group) Company as the pacesetter of the 1993 labor competition.

On January 28, 1994, he was employed by Beijing College of Commerce as a visiting professor for Division I of Beijing College of Commerce.

In January 1994, he was rated as a technical expert by the Beijing Federation of Trade Unions and awarded the title of Patriotic Meritorious Service Pacesetter.

In February 1994, he was rated as an active member of the 1993 Science and Technology Association by Xuanwu District Science and Technology Association.

In June 1994, he participated in the publication of *Chinese Famous Chefs' Specialties (II) Beijing Chapter*, which was compiled by relevant authority of Taiwan.

On October 1, 1994, he was awarded the Senior Chef Qualification Certificate issued by Beijing Municipal Labor Bureau (currently Beijing Municipal Bureau of Human Resources and Social Security).

From 1994 to 1995, he served as Deputy Manager and Chef of Taifeng Restaurant.

On February 6, 1995, he obtained the Qualification Certificate for Chinese Cooking Technician.

In March 1995, he was awarded the title of Advanced Individual in the 1994 Xuanwu District Science and Technology Association by the Xuanwu District Science and Technology Association of Beijing.

In March 1995, he was selected as the 1994 Outstanding Communist Party Member by Beijing Xiangda Catering (Group) Company.

In May 1995, he obtained the Certificate of Senior Technician in Chinese Cooking issued by the Ministry of Labor of the People's Republic of China (currently Beijing Municipal Bureau of Human Resources and Social Security).

On May 25, 1995, he was employed as a member of the Cooking Professional Evaluation Committee of the Workers' Evaluation Committee of the Organs under Central Jurisdiction.

In July 1995, he was awarded by the Xuanwu District People's Government of Beijing the title of the Second Group of Outstanding Scientific, Educational, Technical & Management Talents.

In 1995, he took a temporary position as Chef of Mimisai in Tokyo, Japan while serving Deputy Manager of Taifeng Restaurant. He returned to China in 1996.

On February 26, 1996, he was named the 1995 Outstanding Communist Party Member of Beijing Xiangda Catering (Group) Company.

In December 1996, he was named Beijing Patriotic Health Advanced Worker by Beijing Patriotic Health Campaign Committee.

In March 1997, he was named the No. 1 Scholar of the Cooking Industry in Xuanwu District in 1996.

On July 17, 1997, he was appointed as a member of the Beijing Famous Dishes Appraisal Committee to celebrate Hong Kong's Return to China.

In 1997, he assumed the Technical Consultant of Sunworld Dynasty Hotel Beijing.

On December 26, 1997, he joined Beijing Chinese Famous Chef Association.

In January 1998, he was named Beijing Patriotic Health Advanced Worker by Beijing Patriotic Health Campaign Committee.

In 1998, he participated in the drafting of standards for Chinese culinary masters convened by the Ministry of Internal Trade (currently Ministry of Commerce of the People's Republic of China).

On May 13, 1998, he was employed as a judge of Beijing Catering Industry Workers and Technicians Evaluation Committee.

On August 31, 1998, he was employed as the Technical Consultant of Fangshan Restaurant.

In September 1998, he was employed as Chief Judge of Beijing Fish Culture in Running Water Grand Prix.

In September 1998, he was awarded the Certificate of Vocational Skill Appraisal Appraiser issued by Beijing Labor Bureau (currently Beijing Municipal Bureau of Human Resources and Social Security).

On November 24, 1998, he led a delegation to perform at Metropark Hotel Kowloon in Hong Kong. Hong Kong's 23 news media rushed to report, causing a sensation in Hong Kong.

In March 1999, he cooked for an important and special banquet for more than 10 shareholders from Canada, Hong Kong and the Mainland at the Sunworld Dynasty Hotel.

On April 6, 1999, he was appointed as the judge of the 3rd Beijing Cooking Service Technology Competition.

On June 30, 1999, he obtained the junior graduation certificate of the Correspondence College, Party School of the Central Committee of CPC.

On June 30, 1999, he joined Beijing Cuisine Association.

In September 1999, he was recognized as a national first-class judge of cooking by the Ministry of Internal Trade (currently Ministry of Commerce of the People's Republic of China).

In September 1999, he was employed by the Beijing Municipal Commerce Commission as the "Judge of the First Yanjing Beer Cup Popular Food Festival in Beijing to Celebrate the Motherland".

On September 29, 1999, he personally presided over a large-scale banquet for 500 people - Jia Qinglin granted Henry Fok and Li Ka-shing the title of Honorary Citizens of Beijing.

In October 1999, he assumed the chief judge of the 1st Cooking Competition of Sunworld Dynasty Hotel Management Group, in which 38 five-star hotels participated.

On November 12, 1999, he was employed as a judge of hot dish in the 4th National Cooking Competition.

In 1999, he was granted a pure gold five-star medal by the president of the Sunworld Dynasty Hotel Management Group.

On February 13, 2000, he attended the recording of Spring Festival Program Showing Unique Cooking Skills by Four Cooking Masters from Hong Kong, Guangdong, Taiwan and Chinese mainland. Master Li Qigui cooked the Nine Dragon Eight Treasure Tripod on the spot.

In 2000, he celebrated the Spring Festival with famous actors Xie Yuan, Li Qi and Yin Xiangjie on TV.

In 2000, he was invited to attend the Norwegian Embassy's Thriving – Performing Unique Cooking Skills.

In 2000, he was invited to cook a prime minister's banquet for the Norwegian prime minister in Shanghai Jinjiang Hotel.

In March 2000, he took group photos with Zhang Shiyao, President of the World Cooking Federation, President Yang Liu and Secretary General Lin Zepu at the award ceremony in Tokyo, Japan, and with two winners of the gold medals respectively in the individual and team competition.

On March 6, 2000, he took part in the 3rd World Championship of Chinese Cuisine held in Tokyo, Japan, serving as the chief leader of all three teams from Beijing (Beijing Quanjude, Beijing Gongdefu and Beijing Dadong Roast Duck Restaurant).

On March 9, 2000, he was appointed by the Tokyo Organizing Committee of the 3rd World Championship of Chinese Cuisine as a member of the Evaluation Committee of the 3rd World Championship of Chinese Cuisine.

On March 30, 2000, he was awarded the title of Chinese Cooking Master by the Ministry of Internal Trade (currently Ministry of Commerce of the People's Republic of China).

In June 2000, he was employed as a judge of the Beijing Tourism Group's Youth Skills Competition.

In July 2000, Mr. Chai Zemin, Chinese ambassador to the United States, wrote an inscription for Master Li Qigui, "Build the Brand of Sunworld Dynasty Hotel and Create First-class Achievements".

On August 1, 2000, he was employed by Beijing Cuisine Association as an appraiser of famous Beijing-style famous dishes.

On August 8, 2000, he was employed as the executive director by Beijing Dongfang Meichu Food Culture Exchange Center.

On September 22, 2000, he took a photo with the leaders of the Beijing Municipal Commerce Commission and the national judges and masters of the catering service industry.

In October 2000, he was a judge on Beijing TV with Cao Wei, Ying Zhuang and American Zhong Sheng.

In October 2000, he was employed as the Technical Consultant by Beijing Gongdefu Catering Management Co., Ltd. and celebrated the opening ceremony of Beijing Longxiang Restaurant with famous national actors.

On October 26, 2000, he was invited to Germany to exchange and inspect the unmanned service process of German boutique hotels.

On November 2, 2000, he was rated as China's Top Ten Cooking Masters in 2000 by the Organizing Committee of the 1st Chinese Food Festival.

On November 7, 2000, he took part in a catering delegation of the China Cuisine Association and went to Italy for an inspection.

On November 23, 2000, he joined the La Commanderie des Cordons Bleus de France in Hong Kong.

In November 2000, he cooked at the Sunworld Dynasty Hotel in Beijing for the reception of Sun Fuling, Vice Chairman of the National Committee of the Chinese People's Political Consultative Conference, who came to inspect the work, and introduced the patented dishes and business features.

In 2000, he was selected into the Restaurant Yearbook of China.

On December 24, 2000, he successfully played the role of King of Christmas in the Asia Pacific First Indoor Square of the Sunworld Dynasty Hotel in Beijing and led the princess to celebrate Christmas.

In 2001, he received the visiting ambassadors of France, Mauritius, Mexico and Uruguay and the French President in the Far East, Mr. Guzari, at the Sunworld Dynasty Hotel.

On August 20, 2001, he was awarded the Certificate of the National Advanced Seminar on Hotel Catering Management issued by China Tourist Hotels Association.

On October 12, 2001, he joined the Famous Chefs Professional Committee of the China Cuisine Association (China Famous Chefs Association).

In October 2001, he participated in the compilation of the 1st Selected Dishes by Chinese Culinary Masters & Famous Chefs.

On October 2001, he was invited to Belgium to exchange and inspect the special dishes integrating Chinese and Western styles.

In October 2001, he published a paper on the characteristics of culinary culture that inherits and promotes Beijing's time-honored brands in the Collection of *Essays on Beijing Chinese Famous Chefs*.

In October 2001, he was selected into the 1st Selected Dishes by Chinese Culinary Masters & Famous Chefs.

On November 8, 2001, he was appointed as Chief Judge for the first Norwegian Salmon Pastry Grand Prix.

On June 26, 2002, he was recognized by the World Association of Chinese Cuisine as an international judge of the Chinese Cuisine Technology Competition.

In 2002, as China's Top Ten Culinary Masters, he was granted by French President in the Far East Guzari and Chinese ambassador to France the title of the French Blue Ribbon Badge.

In June 2002, he was awarded a trophy as outstanding international judge by President Zhang Shiyao of the 4th World Championship of Chinese Cuisine.

In October 2002, Sunworld Dynasty Hotel Beijing launched the food festival of Six Famous Chefs Competing in Beijing, causing a sensation across the country.

On June 26, 2002, he was employed by the World Association of Chinese Cuisine as an international judge of the 4th World Championship of Chinese Cuisine, which was held in Kuala Lumpur, Malaysia.

On October 8, 2002, together with the heads of the World Association of Chinese Cuisine and the China Cuisine Association, he participated in the certification and evaluation of the selected dish of the Sunworld Dynasty Hotel, i.e., Ching and Han Royal Dynasty Feast.

In 2002, Minister Jiang Xi wrote an inscription for Master Li Qigui, "Integrate the strengths of all to create unique specialties".

In 2003, he personally presided over the 500-member large-scale conference meal attended by Premier Wen Jiabao of the State Council and the President of the World Bank at the head office of the People's Bank of China, which was well received.

On March 8 and 9, 2003, the 3rd World Championship of Chinese Cuisine was held in Tokyo, Japan. Li Qigui served as the international judge of the Hot Cuisine Group.

In May 2003, he won the qualification of judge for the 5th National Cooking Technology Competition.

In July 2003, the President of the World Association of Chinese Cuisine and the President of the China Cuisine Association wrote an inscription for Master Li Qigui, "Make positive contributions to the promotion of Chinese food culture".

On September 3, 2003, he was invited to Manzhouli to receive new apprentice Liu Zhongxin and performed cooking skills at the Golden Autumn Food Festival.

In September 2003, he was employed as a part-time professor by Beijing Nationality University.

On April 10, 2004, he was appointed as the judge of the finals of the 5th National Cooking Technology Competition.

In August 2004, he cooked for the visiting Spanish ambassador at Sunworld Dynasty Hotel Beijing.

On September 18, 2004, he won the platinum medal with the Batter-fried Butterfly Shrimp with Honey made in the International Famous Chinese Chef's Innovative Dish Exhibition.

In September 2004, he edited the book *Selected Chinese Famous Dishes*.

On November 18, 2004, he was employed by the World Association of Chinese Cuisine as a judge of the International Sculpture Group of the 5th World Championship of Chinese Cuisine.

In 2005, he was selected into the Famous Chef Professional Committee and the Chinese Cooking Association published the book *Chinese Banquets*.

In 2005, he attended an award ceremony of the 5th National High-tech Competition held in the Great Hall of the People.

On June 8, 2005, he participated in the cooking technician teaching program of Beijing Industry and Trade Technicians College.

In 2005, the French Kitchen Emperors awarded Master Li Qigui a five-star platinum medal.

On October 18, 2005, he was recognized by the China Cuisine Association as the first-class judge of Chinese cooking in the national catering industry.

In November 2005, he won the gold medal of the French Kitchen Emperors.

In February 2006, he was accredited as a member of the expert group in the Chinese catering service masters held in Hebei Province.

In March 2006, he was awarded the title of Beijing Super Cooking Master by Beijing Cuisine Association.

On November 28, 2006, he was chef judge of the 4th Sunworld Dynasty Group Cooking Competition.

On December 9, 2006, he participated in the training course for referees and judges of the National Vocational Skills Competition.

In December 2006, he was recognized as a judge of Beijing Catering Special Skills by Beijing Cuisine Association.

On June 17, 2007, he was qualified as a referee in the Vocational Skills Competition of the Vocational Skills Testing Center of the Ministry of Labor and Social Security of the People's Republic of China (currently Ministry of Human Resources and Social Security of the People's Republic of China).

On June 26, 2007, he was rated by the China Cuisine Association as the first-class judge of the national catering industry.

In October 2007, he participated in a large-scale celebration of National Day by Chinese and overseas Chinese in Rotterdam Sea Park, Netherlands.

On October 19, 2007, he was invited to perform cooking skills in Groningen, Netherlands, where the Chinese restaurant Building Beyond Building started business. The Chinese Consul in the Netherlands and the Mayor of Groningen attended the event.

On November 27, 2007, he presided over the 2nd Grand Prix of the Sunworld Dynasty Hotel Management Group and served as the chief referee.

In 2007, he was invited to Europe to perform the Chinese Tripod Banquet and exchange Chinese and Western food culture and cooking skills with gourmets.

On October 21, 2008, he presided over a large banquet with 146 tables for celebrating 50 Anniversary of Beijing Gas Group.

On August 9, 2008, he cooked a Chinese tripod banquet for former German Chancellor Schröder at the welcoming ceremony of the Centennial Olympics.

On August 12, 2008, he was awarded the Special Contribution Award by the Norwegian Seafood Agency.

In October 2008, he was named "International Master of Chinese Food" by the World Association of Chinese Cuisine.

In June 2008, he was included as member of the 4th Council of the Beijing Cuisine Association at the 5th meeting.

On September 16, 2008, the Ching and Han Royal Dynasty Feast of the Sunworld Dynasty Hotel he presided over was selected into the *Appreciation of Chinese Menu* compiled by the Expert Working Committee of the China Cuisine Association.

On October 16, 2008, he was employed by the World Association of Chinese Cuisine as the international judge of the 6th World Championship of Chinese Cuisine and served as the leader of the cold dish group.

In April 2009, he was employed by the China Cuisine Association as a judge of the 6th National Cooking Competition.

On October 20, 2009, he joined the Famous Chefs Committee of the World Association of Chinese Cuisine.

In October 2009, he was selected as the first member of the Professional Committee of the World Federation of Chinese Cuisine.

On August 27, 2009, he won the "Study Certificate" of Taiwan's Food Culture.

In 2010, he participated in the compilation of the book *Chinese Chefs Abroad*.

In 2010, he obtained the Patent Certificate for Utility Model of China's Eight Treasure Tripod.

On April 20, 2010, he obtained the patent of Roast Duck with Ganoderma Lucidum.

In April 2010, he went to the Central Military Commission to train chefs, giving lectures and performance on cooking, balancing diet and nutrition.

On February 4, 2010, he cooked the Chinese Eight Treasures Tripod Banquet for 175 ambassadors to welcome the Spring Festival.

In June 2010, he compiled the Essence Album of Works by *Chinese Cooking Masters*.

On August 26, 2011, he held a patent technology inheritance ceremony and 60th birthday with the presence of Zhang Shiyao, Su Qiucheng and Yang Dengyan.

In 2012, he cooked a famous patent dish Chinese Eight Treasures Tripod Banquet for former Australian Prime Minister Hawke in Manjuji.

On February 13, 2012, led by the head of the team, he successfully performed Snowflake Longxu Noodles and Chinese Eight Treasures Tripod in New York, USA. Former US Labor Secretary Mo Tiancheng, Counselor of the Chinese Consulate in the United States Xu Bing, US Tourism Director Robert, Florida government officials and more than 200 people from various states attended.

In April 2012, he won the Golden Jazz Award for Chinese Cooking Masters of the China Cuisine Association.

On April 17, 2012, he was appointed by Singapore as Vice Chairman of the Imperial Challenge Judges Panel.

In 2012, he won the gold medal for the group in the 7th World Championship of Chinese Cuisine.

In August 2012, he was appointed by the World Association of Chinese Cuisine as a visiting lecturer for the World Association of Chinese Cuisine's International Advanced Seminar on Chinese Cuisine.

In 2012, he won the gold medal for group in the 7th World Championship of Chinese Cuisine at Inner Mongolia Erdos Yitai Kairuibo International Hotel, Sunworld Dynasty Hotel Management Group.

On December 6, 2012, he restored Hongqinglou Restaurant, one of Beijing's eight famous restaurants.

In 2012, he was employed as an international judge by the World Association of Chinese Cuisine.

In September 2012, he completed the 6th International Judges Training Course of the World Association of Chinese Cuisine and won the Judges Certificate.

On November 16, 2012, he was appointed as the judge of the 7th World Championship of Chinese Cuisine.

In 2012, he won the Gold Medal Judge Award in the 7th World Championship of Chinese Cuisine.

In 2013, he was appointed as Chief Judge of Beautiful China's Innovative Cuisine.

On August 28, 2013, he was employed by the Sunworld Dynasty Hotel Management Group as the chief consultant of kitchen operation.

In September 2013, he performed at the Czech and Slovak Food Festival held by the China Cuisine Association.

On November 11, 2013, he was appointed by the China Cuisine Association as the head of the hot dish evaluation team in Kunming, the largest competition area in the country, in the 7th National Cooking Competition.

In 2013, he served as the judge of the China Hotel Industry Famous Chefs and Famous Dishes Competition.

In 2013, he was employed as Consultant of the Catering Committee of China Hotel Purchasing & Supplying Association.

In September 2013, he was employed as Consultant of Longquan Hotel at a large-scale international event held in Beijing, i.e., China Division of the Beijing Marathon Rally.

In 2014, he was employed as an expert judge of Delicacies Fresh and Fabulous Flavors.

In 2014, he won the Outstanding Referee Award at Metro Cup Championship.

From 2013 to 2018, he guided, trained and evaluated the chef team for Zhongnanhai Service Office.

On December 17, 2014, he was awarded the "Catering Culture Merit Award" by the China Food Culture Research Association.

On January 31, 2015, he won the 2014 Special Contribution Award for Dissemination of Food Culture selected by 8 associations.

On August 7, 2015, he was registered at the China Professional Talent Evaluation Expert Committee as a senior hotel manager in the China Professional Talent Pool.

In September 2015, he was recognized as a senior registered Chinese culinary master by the China Cuisine Association.

In May 2016, he conducted an upgrade evaluation of senior chefs from the Organization Department of the CPC Central Committee.

In June 2016, he was included in a book commemorating the 25th anniversary of the World Association of Chinese Cuisine.

In June 2016, he was awarded by the World Association of Chinese Cuisine the title of Fenghuo Cup Chinese Chef Artist.

On March 26, 2017, Jiang Junxian, President of the China Cuisine Association, presided over the ceremony of receiving apprentices to carry forward Master Li Qigui's cooking skills, spreading the spirit of craftsman.

In May 2017, he was selected by the China Cuisine Association as a 30-year meritorious figure in Chinese catering.

On August 3, 2017, he took part in the evaluation of Yellow River Ecological Food Cities Weinan City and Luoyang by the Expert Group of China Cuisine Association.

On April 9, 2018, he was employed by the China Cuisine Association as a judge to evaluate 138 registered Chinese culinary masters in Dalian.

In May 2018, he was awarded by the China Cuisine Association the title of outstanding contributions to the inheritance of Chinese catering industry skills in the 40 years of Reform and Opening Up.

In October 2018, with the approval of the Steering Committee for the Inheritance of Master and Apprentice of the Hall of Fame for Chinese Cooking Masters, Li Qigui, a senior registered Chinese cooking master, was named the mentor of the Hall of Fame for Chinese Culinary Masters.

On October 20, 2018, he was elected as the chief judge of the cold dish group of the 6th World Championship of Chinese Cuisine.

On December 25, 2018, he was granted by President Jiang Junxian of the China Cuisine Association with the certificate to the Chinese Eight Treasures Tripod Banquet and wrote the inscription, "Special Chinese Eight Treasures Tripod Banquet".

On November 5, 2018, he won the A-level Certificate of Registered Referee in the Vocational Skills Competition of China Cuisine Association.

On November 22, 2018, he was appointed as a judge of the China Cuisine Association to evaluate 75 registered cooking masters in Changchun City, Jilin Province.

On July 2, 2019, he was appointed as the head of the hot dish evaluation team for Beijing Area in the 8th National Cooking Competition organized by the China Cuisine Association.

On September 24, 2019, he was appointed as the head of the judging team for the finals of the National Cooking Competition in Tianjin organized by the China Cuisine Association.

2013年，李启贵大师参加中国烹饪协会举办的营口市"红运杯"营菜大赛，任评委。
In 2013, Li Qigui served as a judge in the Yingkou "Red Cloud Cup" Competition held by the China Cuisine Association.

2019年，李启贵大师和马德明恩师的女儿，原宣武区委副书记郑文齐合影。
In 2019, Li Qigui and Zheng Wenqi, daughter of his mentor Ma Deming and former deputy secretary of Xuanwu District Committee, posed for a photo.

2019年，李启贵首都餐饮文化研究发展基地成立仪式。
In 2019, the inauguration ceremony of the Beijing Restaurant Culture Research and Development Base.

李迹周书雅集

李迹周书雅集。
A gathering of this book creators.

2019年9月，在天津举行第八届全国烹饪大赛总决赛，任评判组组长。
In September 2019, Li Qigui was the head of the judging panel at the finals of the 8th National Culinary Competition held in Tianjin.

2012年，第七届中国烹饪世界大赛在新加坡举行，李启贵大师任本届大赛评委。
In 2012, the 7th World Championship of Chinese Cuisine was held in Singapore, and Li Qigui served as a judge of this competition.

2013年，美丽中国——第四届中国饭店业名师名菜创新大赛。
In 2013, beautiful China: the 4th China Restaurant Industry Master Cuisine Innovation Competition.

2016年1月30日，"聚国厨典范，传技艺精粹。北京烹饪协会春节团拜会"合影。
On January 30, 2016, group photo of Beijing Culinary Association's Spring Festival Gathering.

北京市商委领导与餐饮服务业国家级评委、大师名师合影。
Group photo of leaders of Beijing Municipal Commission of Commerce with national judges and masters of food service industry.

2013年,李启贵大师被聘为美丽中国创新大赛总裁判长。
In 2013, Li Qigui was appointed as the Chief Judge of the Beautiful China Innovation Competition.

2019年,北京烹饪协会春节团拜会暨新书首发式合影。
In 2019, Beijing Culinary Association Spring Festival Reunion Party and New Book Premiere.

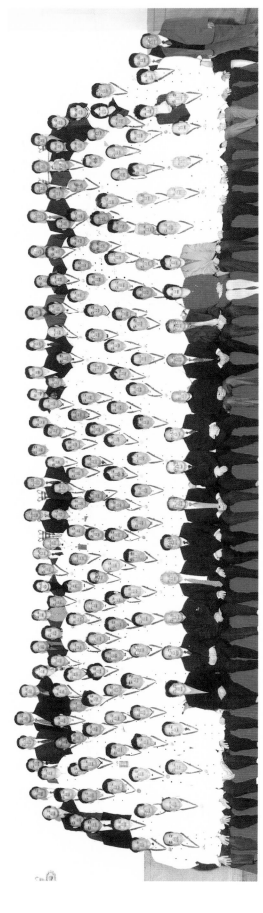

2001年10月12日,在人民大会堂出席中国烹饪协会名厨专业委员会(中国名厨联谊会)成立大会。
On October 12, 2001, the inaugural meeting of the Famous Chefs Professional Committee of China Culinary Association (China Famous Chefs Association) was held at the Great Hall of the People in Beijing.

2011年8月26日,中国十佳烹饪大师李启贵60大寿庆典暨收徒、专利技术传承仪式合影。
On August 26, 2011, Li Qigui, one of China's top 10 culinary masters, celebrated his 60th birthday and took on apprentices to pass on his patented skills.

2012 年 7 月，餐饮业国家职业技能鉴定专家委员会工作会议合影。
In July 2012, group photo of the working meeting of the Expert Committee on National Vocational Skills Certification in Catering Industry.

中国烹饪协会 30 周年庆典大会上，"中国餐饮 30 年功勋人物奖" 获奖人员合影。
Li Qigui received the award of 30 years of meritorious person of Chinese catering in the 30th anniversary celebration of China Cuisine Association.